Sustainable Municipal Solid Waste Management: A Local Issue with Global Impacts

Sustainable Municipal Solid Waste Management: A Local Issue with Global Impacts

Editors

Hani Abu-Qdais
Anna Kurbatova

MDPI • Basel • Beijing • Wuhan • Barcelona • Belgrade • Manchester • Tokyo • Cluj • Tianjin

Editors
Hani Abu-Qdais
Jordan University of Science
and Technology
Jordan

Anna Kurbatova
Peoples' Friendship
University of Russia (RUDN
University)
Russia

Editorial Office
MDPI
St. Alban-Anlage 66
4052 Basel, Switzerland

This is a reprint of articles from the Special Issue published online in the open access journal *Sustainability* (ISSN 2071-1050) (available at: https://www.mdpi.com/journal/sustainability/special_issues/sustainable_municipal_solid_waste_management).

For citation purposes, cite each article independently as indicated on the article page online and as indicated below:

LastName, A.A.; LastName, B.B.; LastName, C.C. Article Title. *Journal Name* **Year**, *Volume Number*, Page Range.

ISBN 978-3-0365-5681-9 (Hbk)
ISBN 978-3-0365-5682-6 (PDF)

© 2022 by the authors. Articles in this book are Open Access and distributed under the Creative Commons Attribution (CC BY) license, which allows users to download, copy and build upon published articles, as long as the author and publisher are properly credited, which ensures maximum dissemination and a wider impact of our publications.

The book as a whole is distributed by MDPI under the terms and conditions of the Creative Commons license CC BY-NC-ND.

Contents

About the Editors vii

Hani A. Abu-Qdais and Anna I. Kurbatova
Editorial: Sustainable Municipal Solid Waste Management: A Local Issue with Global Impacts
Reprinted from: *Sustainability* **2022**, *14*, 11438, doi:10.3390/su141811438 1

Hani A. Abu-Qdais and Anna I. Kurbatova
The Role of Eco-Industrial Parks in Promoting Circular Economy in Russia: A Life Cycle Approach
Reprinted from: *Sustainability* **2022**, *14*, 3893, doi:10.3390/su14073893 5

Safwat Hemidat, Ouafa Achouri, Loubna El Fels, Sherien Elagroudy, Mohamed Hafidi, Benabbas Chaouki, Mostafa Ahmed, Isla Hodgkinson and Jinyang Guo
Solid Waste Management in the Context of a Circular Economy in the MENA Region
Reprinted from: *Sustainability* **2022**, *14*, 480, doi:10.3390/su14010480 21

Guilberto Borongan and Anchana NaRanong
Practical Challenges and Opportunities for Marine Plastic Litter Reduction in Manila: A Structural Equation Modeling
Reprinted from: *Sustainability* **2022**, *14*, 6128, doi:10.3390/su14106128 45

Esra'a Amin Al-Athamin, Safwat Hemidat, Husam Al-Hamaiedeh, Salah H. Aljbour, Tayel El-Hasan and Abdallah Nassour
A Techno-Economic Analysis of Sustainable Material Recovery Facilities: The Case of Al-Karak Solid Waste Sorting Plant, Jordan
Reprinted from: *Sustainability* **2021**, *13*, 13043, doi:10.3390/su132313043 75

Olesya A. Buryakovskaya, Anna I. Kurbatova, Mikhail S. Vlaskin, George E. Valyano, Anatoly V. Grigorenko, Grayr N. Ambaryan and Aleksandr O. Dudoladov
Waste to Hydrogen: Elaboration of Hydroreactive Materials from Magnesium-Aluminum Scrap
Reprinted from: *Sustainability* **2022**, *14*, 4496, doi:10.3390/su14084496 89

Aseel Najeeb Ajaweed, Fikrat M. Hassan and Nadhem H. Hyder
Evaluation of Physio-Chemical Characteristics of Bio Fertilizer Produced from Organic Solid Waste Using Composting Bins
Reprinted from: *Sustainability* **2022**, *14*, 4738, doi:10.3390/su14084738 123

Camilo Venegas, Andrea C. Sánchez-Alfonso, Fidson-Juarismy Vesga, Alison Martín, Crispín Celis-Zambrano and Mauricio González Mendez
Identification and Evaluation of Determining Factors and Actors in the Management and Use of Biosolids through Prospective Analysis (MicMac and Mactor) and Social Networks
Reprinted from: *Sustainability* **2022**, *14*, 6840, doi:10.3390/su14116840 135

Feriel Kheira Kebaili, Amel Baziz-Berkani, Hani Amir Aouissi, Florin-Constantin Mihai, Moustafa Houda, Mostefa Ababsa, Marc Azab, Alexandru-Ionut Petrisor and Christine Fürst
Characterization and Planning of Household Waste Management: A Case Study from the MENA Region
Reprinted from: *Sustainability* **2022**, *14*, 5461, doi:10.3390/su14095461 159

About the Editors

Hani Abu-Qdais

Hani Abu-Qdais is a Professor of Environmental Engineering in the Civil Engineering Department of Jordan University of Science and Technology (JUST). He has served in several academic positions, including Assistant to the University President, Dean of Scientific Research and Graduate Studies, and Director of Queen RANIA Centre for Environment. Having more than 30 years of diverse experience in public, private, and academic sectors, Prof. Abu-Qdais has obtained more than USD 3 million in funds for his research and consultation projects. He has published more than 60 papers in international refereed journals and proceedings of international conferences (H Index: Scopus = 21, Google Scholar = 27). Most of his publications focus on integrated solid waste management, circular economy, waste to energy, climate change, and water treatment and management. In 2021, Professor Abu Qdais was listed among the World's top 2% of scientists by Stanford University. Furthermore, he is a winner of the Shoman Award for Arab scientists in 2015. Prof. Abu Qdais has served as a consultant for several national and international agencies, such as the Ministry of Local Administration, Ministry of Environment, UNDP, GIZ, USAID, WHO, IDRC, FAO, SIDA, EU, and projects funded by the World Bank.

Anna Kurbatova

Dr. Anna I Kurbatova is an Associate Professor in the Department of Environmental Safety and Product Quality Management, Institute of Environmental Engineering, Peoples' Friendship University of Russia (RUDN University). Currently, she serves as the Deputy Head of the Greenhouse Gas Quality Validation and Verification Body at RUDN University, Institute of Environmental Engineering, and the Head of the WtERT branch in Russia. Furthermore, she is a Member of the Committee on Extended Producer Responsibility, Business Russia. Dr. Kurbatova obtained her master's degree in 2002 from the Faculty of Physics, Mathematics and Natural Sciences at RUDN University, with a specialty in "chemistry", and a focus in the fields of inorganic chemistry and analytical chemistry. In 2006, she completed her PhD studies in the Department of Ecological Monitoring and Forecasting at the PFUR and defended her PhD thesis in environmental sciences. Since 2007, Dr. Kurbatova has been Associate Professor and runs several projects in solid waste management and environmental science and technology.

Editorial

Editorial: Sustainable Municipal Solid Waste Management: A Local Issue with Global Impacts

Hani A. Abu-Qdais [1,*] and Anna I. Kurbatova [2,*]

1 Civil Engineering Department, Jordan University of Science and Technology, Irbid P.O. Box 3030, Jordan
2 Department of Environmental Safety and Product Quality Management, Institute of Environmental Engineering, Peoples' Friendship University of Russia (RUDN University), 6 Miklukho-Maklaya Street, 117198 Moscow, Russia
* Correspondence: hqdais@just.edu.jo (H.A.A.-Q.); kurbatova-ai@rudn.ru (A.I.K.)

Citation: Abu-Qdais, H.A.; Kurbatova, A.I. Editorial: Sustainable Municipal Solid Waste Management: A Local Issue with Global Impacts. *Sustainability* **2022**, *14*, 11438. https://doi.org/10.3390/su141811438

Received: 5 September 2022
Accepted: 8 September 2022
Published: 13 September 2022

Publisher's Note: MDPI stays neutral with regard to jurisdictional claims in published maps and institutional affiliations.

Copyright: © 2022 by the authors. Licensee MDPI, Basel, Switzerland. This article is an open access article distributed under the terms and conditions of the Creative Commons Attribution (CC BY) license (https://creativecommons.org/licenses/by/4.0/).

On a global level, communities are generating and disposing of increasing quantities of solid waste. This waste includes different categories like municipal solid waste [1], industrial solid waste [2] and agricultural waste [3]. Unless managed in a sustainable manner, such wastes will lead to public health risks, adverse environmental impacts and other socio-economic problems [4]. Therefore, regulations should be enforced to encourage sustainable practices like reduction, reuse and recycling of solid waste.

This special issue of Sustainability provides an overview of sustainable solid waste management by publishing articles on recent research concepts, models and case studies, in order to stimulate the scientific potential of this field and for a better understanding of the circular economy principles. To achieve that, specialists from a range of disciplines, practices and sectors were invited to contribute their research findings. The number and diversity of submissions that we received to the call from different regions of the world indicate the growing importance of sustainable solid waste management. The special issue consists of eight articles. Each of them contributes to the knowledge in the field by presenting the features of cutting-edge treatment technologies and the latest modeling techniques. From these eight articles, several unique lessons have emerged.

The first article [5] tackled the issue of sustainable management and the use of biosolids in Columbia. By adopting the Prospective Analysis (MicMac and Mactor) and Social Networks approach, the researchers managed to analyze those factors that influence the management and use of biosolids. The application of such methodology that was focused on biosolids allowed for the prioritization of determinants, the evaluation of the level of involvement and communication between actors and other aspects that have not been considered previously in the management of wastewater treatment plants in Colombia. The study concluded that there is a weak linkage between the stakeholders which requires work to improve the communication skills between them.

The challenge of marine plastic littering in Manila, Philippines was addressed in the second article [6]. The researchers used structural equation modeling to examine the practical challenges and opportunities to influence the reduction of marine plastic littering. By using an online survey, the investigators constructed the model which had later been subjected to validation through interviews and focused group discussions, where the developed model had been validated to a good internal consistency. The findings revealed that environmental governance, in terms of waste management policies and guidelines, COVID-19 regulations for waste management, community participation and socio-economic activities had a positive impact on decreasing the amount of marine plastic littering.

A case study to characterize and plan household waste in Algeria was covered in the third article [7]. To assess the spatial distribution of the waste and potential drivers of waste production, the study adopted a geospatial analysis by integrating the principal component analysis with GIS. The results showed that household waste management is influenced by factors related to the size of the settlement and the characteristics of waste management companies. The study indicated that the urban waste generation rate is estimated to be at 0.8 kg·inhab·day^{-1}, while the recycling rate is around 7%, composting 1% and the rest of municipal waste flow is disposed of in sanitary landfills or dumpsites. The investigators concluded that a combination of multiple regression analysis and principal component analysis is efficient to describe and understand waste production.

The fourth article [8] dealt with the composting process by converting the organic fraction of municipal solid waste to mature compost using the composting bin method. Several materials, including animal waste, agricultural waste and humic substances, were added to the municipal waste and subjected to composting. Temperature, pH, EC, organic matter (OM percent), the C/N ratio and micronutrients (N, P, K) were monitored on a weekly basis. The study concluded that bin composting may be used as a quick check to ensure that the output of the long-term public projects will meet the sustainability requirement, enhance the ecosystem services and will mitigate the climate change impact that may be caused by the disposal of the solid waste in the hot climate regions.

Waste to Hydrogen energy was the topic which was covered by the fifth article [9]. The researchers examined the production of hydrogen energy by processing an aluminum scrap. Ball-milled hydro-reactive powders of Mg-Al scrap with 20 wt.%, with and without additives were prepared. Their hydrogen yields and reaction rates in a 3.5 wt.% NaCl aqueous solution at 15–35 °C were estimated and compared. The results showed that samples with 20 wt.% Wood's alloy and with no additives demonstrated the highest hydrogen yields of (73.5 ± 10.0)% and (70.6 ± 2.5)%, respectively. However, their maximum reaction rates were the lowest. On the other hand, the results indicated that the variation in reaction kinetics was attributed to the difference in the scrap powder particle size, where the samples with a salt additive had the finest particle size and the fastest reaction kinetics at the beginning of the reaction.

An interesting topic is addressed in the sixth article [10]. A case study has been presented to demonstrate the role of Eco-Industrial parks in promoting the circular economy in Russia. Given the fact that the circular economy is one of the priorities of the country's economy, the researchers adopted the life cycle assessment as a tool to compare two scenarios of solid waste management. The first scenario is the disposal of the solid waste at the landfill, while the second one is to divert the generated solid waste from landfills to an eco-industrial park, where it is subjected to processing for materials and energy recovery. The life cycle analysis showed that diverting 1.813 million tons of mixed municipal solid waste that is generated in Moscow to EIP would lead to a reduction in environmental impacts. The total global warming potential of the EIP scenario is less, by 59%, than the direct landfilling scenario, while the eutrophication, acidification, smog and ozone depletion are less, and fossil fuel depletion impacts under the second scenario are less, by 81%, 26%, 18% and 81%, respectively. The study indicated that the adoption of a circular economy in Russia is still in its early stages. To create an enabling environment for the promotion of a circular economy, the study recommended overcoming several institutional, technical and social barriers, where the Russian higher educational institutions can play a major role in overcoming such obstacles.

Article seven [11] was devoted to solid waste management in the MENA region in the context of a circular economy. The paper presents a comprehensive overview of the national municipal solid waste systems of 10 countries in the MENA region with a detailed evaluation of four countries (Jordan, Egypt, Algeria and Morocco). Current practices of the solid waste management adopted and the approaches followed by each country to integrate the circular economy principles in their solid waste management were identified. The researchers indicated that the solid waste management in the studied countries is still disposal driven, where it ranges from 73% in the UAE up to 97% in Libya. Furthermore, the organic fraction constitutes the highest percentage of the solid waste stream that ranges from 504% in Algeria to 70% in Morocco. Full cost recovery of the solid waste services does not exist in any of the studied countries. Furthermore, the existing legal and administrative frameworks do not enable the adoption of a circular economy. Therefore, the study recommends that the governments should promote an integrated solid waste management hierarchy and set up a national policy regarding the diversion of waste from landfill by making a shift from waste disposal toward waste management and the circular economy.

Finally, the last article [12] conducted a techno-economic analysis of the Al-Karak solid waste sorting plant in Jordan. The article investigated the possible technical and economic performance of the sorting plant in order to achieve financial sustainability and increase profits to cover its operating costs. Possible different equipment and material flow through the plant were proposed. To assess the feasibility of the proposed options, an economic model was used based on three economic factors, which are net present worth (NPW), return on investment (ROI) and payback period values. The results showed that the input materials contain a high fraction of recyclable materials like paper, cardboard, plastic and metals, which account for 63% of the solid waste stream. The analysis shows that to be economically feasible, the plant operation should be on a two or three shifts basis, where the rate of return will be 3.5 and 4.4 with a payback period of 3 and 2 years, respectively.

Funding: This research received no external funding.

Conflicts of Interest: The authors declare no conflict of interest.

References

1. Vergara, S.E.; Tchobanoglous, G. Municipal Solid Waste and the Environment: A Global Perspective. *Annu. Rev. Environ. Resour.* **2012**, *37*, 277–309. [CrossRef]
2. Csares, M.L.; Ulierte, A.; Mataran, A.; Ramos, A.; Zamorano, M. Solid Industrial waste and their management in Asegra (Granada, Spain). *Waste Manag.* **2005**, *25*, 1075–1082. [CrossRef] [PubMed]
3. Abu Ashoor, J.; Abu Qdais, H.A.; Al Widyan, M. Estimation of Animal and Olive Solid Wastes in Jordan and Their Potential as A Supplementary Energy Source: An Overview. *Renew. Sustain. Energy Rev.* **2010**, *14*, 2227–2231. [CrossRef]
4. Abu Qdais, H.A. Techno Economic Analysis of Municipal Solid Waste Management in Jordan. *Waste Manag.* **2007**, *27*, 1666–1672. [CrossRef] [PubMed]
5. Venegas, C.; Sánchez-Alfonso, A.C.; Vesga, F.-J.; Martín, A.; Celis-Zambrano, C.; González Mendez, M. Identification and Evaluation of Determining Factors and Actors in the Management and Use of Biosolids through Prospective Analysis (MicMac and Mactor) and Social Networks. *Sustainability* **2022**, *14*, 6840. [CrossRef]
6. Borongan, G.; NaRanong, A. Practical Challenges and Opportunities for Marine Plastic Litter Reduction in Manila: A Structural Equation Modeling. *Sustainability* **2022**, *14*, 6128. [CrossRef]
7. Kebaili, F.K.; Baziz-Berkani, A.; Aouissi, H.A.; Mihai, F.-C.; Houda, M.; Ababsa, M.; Azab, M.; Petrisor, A.-I.; Fürst, C. Characterization and Planning of Household Waste Management: A Case Study from the MENA Region. *Sustainability* **2022**, *14*, 5461. [CrossRef]
8. Ajaweed, A.N.; Hassan, F.M.; Hyder, N.H. Evaluation of Physio-Chemical Characteristics of Bio Fertilizer Produced from Organic Solid Waste Using Composting Bins. *Sustainability* **2022**, *14*, 4738. [CrossRef]
9. Buryakovskaya, O.A.; Kurbatova, A.I.; Vlaskin, M.S.; Valyano, G.E.; Grigorenko, A.V.; Ambaryan, G.N.; Dudoladov, A.O. Waste to Hydrogen: Elaboration of Hydroreactive Materials from Magnesium-Aluminum Scrap. *Sustainability* **2022**, *14*, 4496. [CrossRef]
10. Abu-Qdais, H.A.; Kurbatova, A.I. The Role of Eco-Industrial Parks in Promoting Circular Economy in Russia: A Life Cycle Approach. *Sustainability* **2022**, *14*, 3893. [CrossRef]
11. Hemidat, S.; Achouri, O.; El Fels, L.; Elagroudy, S.; Hafidi, M.; Chaouki, B.; Ahmed, M.; Hodgkinson, I.; Guo, J. Solid Waste Management in the Context of a Circular Economy in the MENA Region. *Sustainability* **2022**, *14*, 480. [CrossRef]
12. Al-Athamin, E.A.; Hemidat, S.; Al-Hamaiedeh, H.; Aljbour, S.H.; El-Hasan, T.; Nassour, A. A Techno-Economic Analysis of Sustainable Material Recovery Facilities: The Case of Al-Karak Solid Waste Sorting Plant, Jordan. *Sustainability* **2021**, *13*, 13043. [CrossRef]

Article

The Role of Eco-Industrial Parks in Promoting Circular Economy in Russia: A Life Cycle Approach

Hani A. Abu-Qdais [1] and Anna I. Kurbatova [2,*]

[1] Civil Engineering Department, Jordan University of Science and Technology, Irbid P.O. Box 3030, Jordan; hqdais@just.edu.jo

[2] Department of Environmental Safety and Product Quality Management, Institute of Environmental Engineering, Peoples' Friendship University of Russia (RUDN University), 6 Miklukho-Maklaya Street, 117198 Moscow, Russia

* Correspondence: kurbatova-ai@rudn.ru

Abstract: As an approach to move towards a sustainable waste management system, circular economy (CE) is gaining an increased interest by most countries. Russia is among the countries where the CE is one of the priorities of the country's economy, with a market value of the CE is USD$ 755.05 billion. However, such a strategy is facing challenges and barriers which are country specific. This study aimed to review the status of the CE in Russia and to identify the obstacles that are hindering the country from achieving its objectives. Moreover, the study aimed to evaluate the role of eco-industrial parks (EIP) in Russia in promoting the CE model. The study findings indicate that the CE adoption in Russia is still in its early stages. To create an enabling environment for CE promotion in Russia, there is a need to overcome several institutional, technical, and social barriers. Russian higher educational institutions are playing a major role to create the critical mass of experts that will help the country transition towards a CE model. Using life cycle assessment (LCA) to analyze the environmental performance of one of the EIPs in Russia revealed that such enterprises are more sustainable than the business-as-usual scenarios, under which the generated solid waste is buried into landfill. The comparison shows that by diverting 1.813 million tons of mixed municipal solid waste that is generated in Moscow to EIP would lead to a reduction in environmental impacts. The total global warming potential of the EIP scenario is less, by 59%, than the direct landfilling scenario, while the eutrophication, acidification, smog, and ozone depletion are less, and fossil fuel depletion impacts under the second scenario are less, by 81%, 26%, 18%, and 81%, respectively. Furthermore, the health impacts including carcinogenic, non-carcinogenic, eco-toxicity were found to be 92%, 96%, and 96%, respectively, less than the baseline scenario.

Keywords: circular economy; eco-industrial park; life cycle assessment; solid waste; Russia

Citation: Abu-Qdais, H.A.; Kurbatova, A.I. The Role of Eco-Industrial Parks in Promoting Circular Economy in Russia: A Life Cycle Approach. *Sustainability* 2022, 14, 3893. https://doi.org/10.3390/su14073893

Academic Editor: Donato Morea

Received: 22 January 2022
Accepted: 22 March 2022
Published: 25 March 2022

Publisher's Note: MDPI stays neutral with regard to jurisdictional claims in published maps and institutional affiliations.

Copyright: © 2022 by the authors. Licensee MDPI, Basel, Switzerland. This article is an open access article distributed under the terms and conditions of the Creative Commons Attribution (CC BY) license (https:// creativecommons.org/licenses/by/ 4.0/).

1. Introduction

Since it contributes in achieving sustainable development goals (SDGs), circular economy (CE) is becoming a popular concept that is being promoted by many countries and businesses around the world [1–3]. Circular economy aims to move from a linear economy model, to a more sustainable one in which products, materials, and resources are kept in the system for as long as possible and in which the generation of waste is minimized [4].

In Russia, the promotion of the circular economy is one of the priorities for the Russian economy. The estimated market value of the circular economy in the country is about USD$ 755.05 billion [5]. However, the CE model is still under development [6,7] and studies that deal with CE in Russia are still few [8,9]. Hence, the influence of circular practices on the formation of market requires further research [10]. Recent developments and changes in the waste management framework of the Russian Federation have considered that waste as a resource [11,12]. This has paved the way towards introducing the CE model into the country's economy. Implementation of the concept of circular economy in Russia will not only reduce environmental contamination, but also will lead to acceleration in economic growth and to the creation of new jobs.

EIPs are gaining interest as an approach towards the promotion of CE [13–16], where such projects are building symbiotic relationships between various industries that reside in park areas to achieve waste reuse and recycling and pollution reduction [17]. In many countries, EIPs are emerging, where materials and resources are shared to optimize both economic and environmental performance [18]. For example, in Korea, EIPs have become a central element of Korean industrial innovation strategy to assist in the transition to CE [19].

Considering the fact that the real world-application of CE in EIP is still far from perfect [20] and the scarcity in the research that deals with the measurement and assessment of the circular economy within the eco-industrial parks, there is a pressing need for an evaluation system and approach to test the circular economy potential of such projects [13]. Furthermore, performance measurements of eco-industrial parks are difficult to obtain as the material and energy flows within these facilities are complex; they are available in different forms and measured by different units [21]. One of the issues that needs clarification at the level of industrial symbiosis clusters is the relationship between industrial ecology and circular economy concepts, which indicates the need to address such a gap in knowledge by conducting further research [22]. In addition, the available studies lack characterization of the EIPs' organizational models and analysis of how these models are affecting EIPs' sustainability [23]. Therefore, the issue of EIPs is becoming a research topic in the field of recycling economy [24].

Wenbo (2011) [13] used analytic network process (ANP) to evaluate the circular economy performance of five eco-industrial parks in China. The study found that the ANP method can be effectively applied to circular economy performance evaluation and consequently used in decision making. One of the limitations of such an approach is that the quantified indicator values are largely dependent on the experts' opinion.

Tian et al. (2014) [25] studied the performance of five EIPs in China. A group of ten metrics, including resource consumption, economic development, and waste emissions, were applied in the performance assessment process. The researchers found that absolute energy consumption, fresh water consumption, industrial wastewater generation, and solid waste production in 17 eco-industrial parks had been increased, while the average intensity of the emissions (tons of pollutant per million yuan invested) of the four metrics had been decreased. In addition to economic gain, Wang et al. (2019) [26] assessed the potential environmental impacts of an energy intensive EIP by adopting life cycle assessment (LCA). The results showed that LCA is an effective tool in evaluating environmental impacts. The study found that effective environmental impact reduction could be attained in terms of primary energy, greenhouse emissions, acidification, eutrophication, particulate matter emissions, and human toxicity.

One important issue of EIPs is to measure the environmental sustainability of their operations. An optimal EIP is one which minimizes negative impacts and maximizes positives ones. However, the question is how to measure such sustainability aspects in terms of social, environmental, and economic aspects [27]. Many studies recommended the use of quantitative environmental sustainability indicators. For example, Azapagic and Perdan (2000) [28] used ozone depletion as an environmental indicator, and income distribution as a social indicator, while the value added was used as an economic indicator.

Tools such as life cycle assessment (LCA) have proven to be effective means in assessing the eco-efficiency of industrial parks [28]. Belaud et al. (2019) [29] provided a toolbox for developing an EIP in France by integrating the circular economy concept with life cycle thinking. Zhao et al. (2016) [24] used a multi-criteria decision making approach to develop a framework for assessing the performance of EIP from the perspective of circular economy. By applying the developed model to six EIPs in China, it was found to be an effective tool of assessment.

Recently, several studies used SimaPro software to conduct LCA analysis for assessing the environmental sustainability of various solid waste management options [30–34]. Out of

96 reviewed studies that used LCA in municipal solid waste management, 44 (46%) studies used the SimaPro model [35]. The most commonly used FU in the LCA of MSWM is 1 MT of waste, 88 out of total studies used it as an FU. [35]. However, few LCA studies were conducted in Russia to assess the environmental impacts of solid waste management. For example, Tulokhonova and Ulanova (2013) [36] assessed the environmental impacts of four solid waste management scenarios in Irkutsk using the integrated solid waste management Model. Another study used LCA to assess the environmental impacts of landfills in the Irkutsk region [37]. Kaazke et al. 2013 [38] conducted an LCA study to assess the environmental impacts of solid waste management practices in Khanty-Mansiysk and Surgut as compared to other alternatives. The researchers used the LCA-IWM model to compare various alternatives of solid waste management. Recently, Vinitskaia et al. (2021) [39] used LCA to evaluate the existing and the proposed municipal solid waste management system in Moscow. The study evaluated six scenarios of waste management and found that the largest emissions reduction potential was associated with the refuse derived fuel (RDF) option.

Eco-Industrial Parks in Russia

EIPs will become the basis for Russia's promotion of circular economy [40]. Dorokhina (2018) [12] explored the possibility of adopting the Chinese experience in establishing the EIP in the Russian Federation. According to the Strategy for the Development of Industry for the Processing, Utilization, and Disposal of Industrial and Municipal Waste of the RF until 2030, an eco-industrial park (in Russian eco-techno park) is a united and interdependent complex of industries that utilize materials and energy flow in the process of waste treatment and utilization to manufacture new products, where scientific research and/or educational activities are integral parts of such complexes [15,41]. Thus, from the legislative point of view, an eco-industrial park is an industrial cluster that includes sorting, recycling, and disposal activities of waste within one site [12,42].

In nine regions of the Russian Federation, thermal schemes and regional programs for waste management (including MSW) have been approved. Such programs provide the possibility of the creation of eco-industrial parks. However, none of the regional programs gave a detailed definition and description of the eco-industrial parks; they only denote them as facilities for complex waste processing [41]. As such, this raises many questions on the role of EIP in promoting a CE model in the country.

One major goal of the Russian solid waste strategy is that by the year 2030, about 80% of generated solid waste will be diverted from landfills to recycling facilities as compared with the current level of 10%. The construction and operation of eco-industrial parks in Russia will help in achieving that goal [42]. Therefore, it is planned by the year 2030 to build and operate 70 eco-industrial parks [15]. Figure 1 shows the locations of operating, under construction, and planned EIPs in the Russian Federation.

Currently, eco-industrial parks are in operation in the Perm, Kurgan, Volgograd, Astrakhan, and Rostov regions, while another ten regional eco-industrial parks with a total capacity of more than 2 million tons per year are at various stages of operation, construction, and design in the Southern Federal District [43].

The main objective of this study was to review the status of the transition from a linear to circular economy in Russia and to identify the challenges that are facing such a transition. Furthermore, the potential role of EIPs in promoting the CE concept in Russia was analyzed by assessing the environmental impacts of a model EIP as a case study using life cycle assessment (LCA).

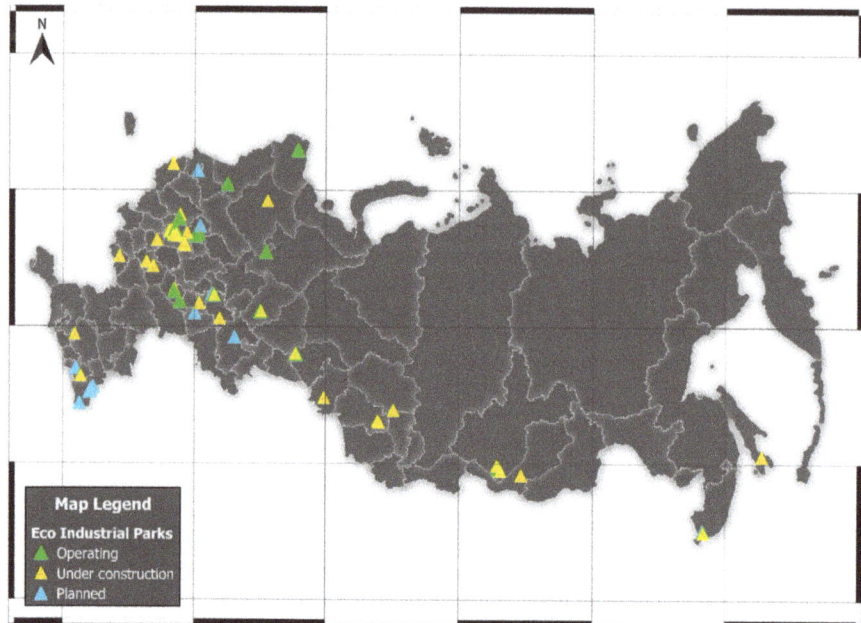

Figure 1. Location of eco-industrial parks in the Russian Federation.

2. Materials and Methods

To achieve the objectives of the study, an extensive literature review on the status of CE both worldwide and in Russia was conducted. To understand the latest progress achieved in moving from a linear to a circular economy, searches within different databases, such as Scopus and web of science, and search engines, such as Google scholar, were carried out. The searches were based on two criteria, namely high impact journals and the relevant key words. As a result, scientific articles, reports, and documents were collected, reviewed, and analyzed. A critical analysis of the CE and the role of EIP in promoting such a model in Russia was carried out. Data on the EIP case study was collected. Information on the amounts and composition of solid waste in Russia was obtained.

2.1. Study Area: Case Study Eco-Industrial Park

For the purpose of this study, one of the EIPs in Russia was selected and its operations subjected to LCA analysis. It is one of the first EIPs in Russia for the sorting, processing, and disposal of municipal solid waste. The total area of the EIP is 1600 hectares. The annual capacity of the EIP is 1.813 million tons of mixed municipal solid waste that is generated in Moscow (the Moscow region generates a total annual amount of 11 million tons). The EIP has two sorting plants that recover recyclables from the mixed solid waste stream. Each plant has an annual capacity of 500,000 tons. The components recovered include ferrous and non-ferrous metals, glass, paper, plastic, and electronic scrap. After sorting, food waste is directed to composting. The residual material, after sorting, is directed to a nearby sanitary landfill, where the produced compost at the EIP territory is applied as a cover on the top of the landfilled waste.

2.2. Life Cycle Assessment

To assess the ecological performance of the case study eco-industrial park, the LCA method is used in compliance with ISO 14040 standard. LCA is defined as a tool to assess the environmental benefits and burdens associated with waste management systems during its life cycle [44]. LCA analysis consists of four interrelated steps as follows:

2.2.1. Goal and Scope Definition

The goal and scope definition step includes the identification of the functional unit as well as the system boundaries. The primary goal of the study was to evaluate the role of the eco industrial parks in Russia in promoting the circular economy concept using a case study EIP. Environmental performance of the EIP was analyzed based on LCA using Simapro 9 Software version [45]. The functional unit (FU) used in the study was 1.813 million tons of municipal solid waste, where all the emissions and impacts were calculated based on this unit. To describe the environmental flows of the system, system boundary was determined. Figures 2 and 3 illustrate the system boundary used in the study which indicates the inputs and outputs of the system. As can be seen, the boundaries of the system were limited to landfill and EIP processes. The collection and transportation activities of the solid waste were excluded from the LCA analysis.

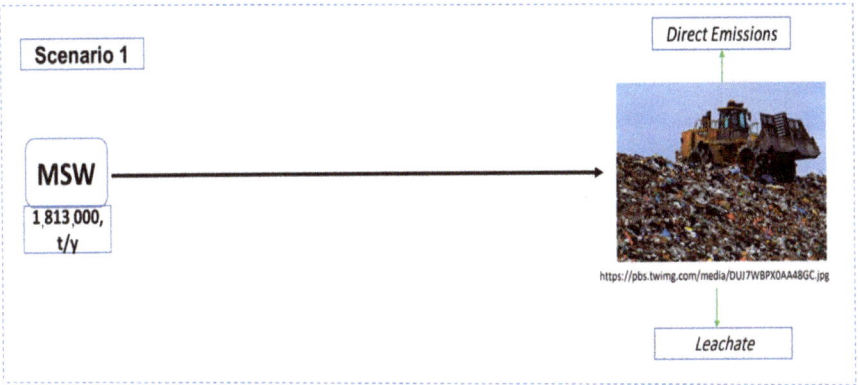

Figure 2. Business as usual scenario where the solid waste is dumped into landfill (scenario 1).

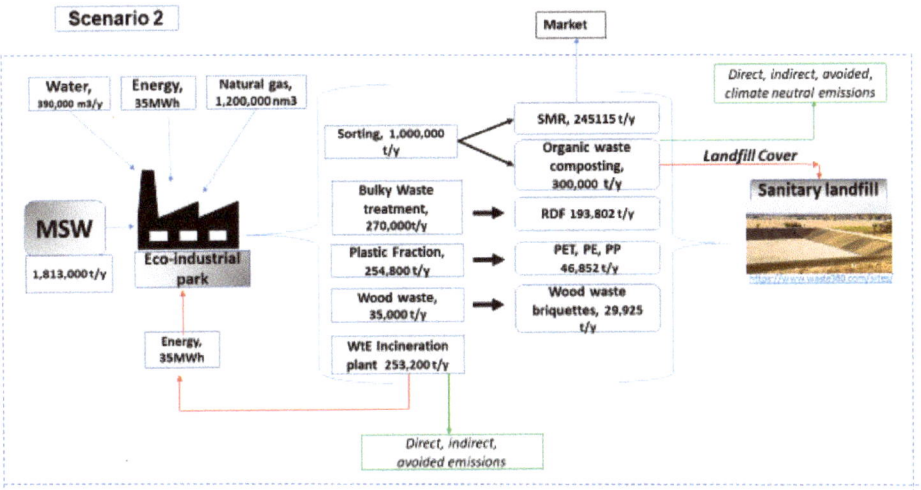

Figure 3. Layout and system boundaries of the eco-industrial park (Scenario 2).

2.2.2. Scenarios

In order to achieve the study objectives, it is important to identify the solid waste management scenarios. The study covers two scenarios, namely direct landfilling of solid waste, which is the main practice that is currently applied in Moscow (business as usual) as shown in Figure 2. Under this scenario, it is assumed that the whole amount of the solid waste (1.83 million tons) is diverted to the landfill. On the other hand, the second scenario is based on the fact that 1.813 million tons per year of municipal solid waste will be diverted from the landfill to the EIP, where it will be subjected to sorting and material and energy recovery. The recyclable items include metals, paper, and plastic. The organic fraction of the waste will be subjected to composting, while the energy recovery happens through incineration and briquettes-making from wood waste (Figure 3). The residual materials, after sorting the mixed solid waste, will be disposed of in a landfill.

2.2.3. Life Cycle Inventory

Life cycle inventory analysis identifies the list of materials as well as energy input and output. The data on the input and output were obtained from the documents published on the web about the EIP. This includes the annual amount of solid waste received on the facility from Moscow, the energy and water needed to run the facility, and the amounts of waste directed to recycling, energy recovery, and composting. Furthermore, the data on the recovered product types and amounts were collected. Table 1 shows the inputs and outputs from the EIP, while the composition of the solid waste generated in Moscow is shown in Figure 4. The compiled data from the LC inventory were introduced into SimaPro software version 9 [45]. Since the data on the waste quantities are available, and the waste treatment processes and recovery are well established, the cut-off approach was used in determining the level of environmental impacts [46]. The cut-off level used in the assessment was 1%. The main characteristics and assumptions for the LCA of the EIP processes are presented in Table 2.

Table 1. Materials and resources Input and output of the case study EIP.

	Input	Output
Solid waste quantity	1,813,000 ton/year	101,824 tons/year to landfill
Sorting	1,000,000 ton/year	Material recovery from sorting 245,115 t/y. This includes - Paper, cardboard 133,260.0 ton/year - Black metal 24,792.3 ton/year - Non-ferrous metal 11,598.3 ton/year - Glass 33,060.0 ton/year - Plastics (MIX) 42,404.8 ton/year. - Compost 300,000 ton/year
Composting (25% food of waste)	453,250 ton/year	300,000 ton/year
From bulky to RDF	270,000 ton/year	RDF, 193,802 t/y
WtE plant	253,200 ton/year	Energy, 35 Mwh
Plastic fraction	254,800 ton/year	PET, PE, PP 46,852 ton/year
Water	390,000 m^3	
Electric Energy	35 Mwh	
Natural gas	1,200,000 nm^3	

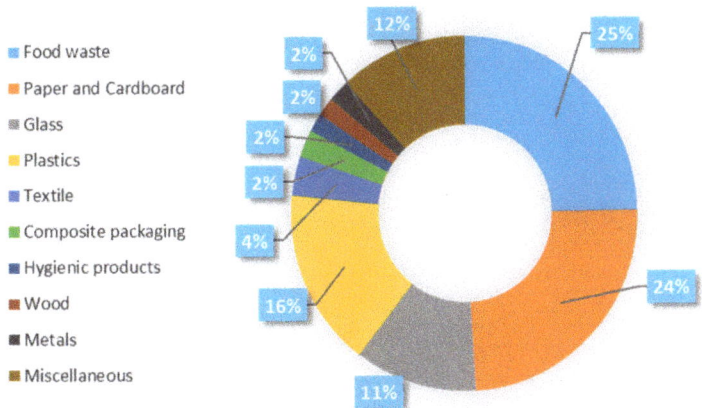

Figure 4. Physical composition of solid waste generated in Moscow Region [47].

Table 2. Main characteristics and assumptions of the EIP processes.

Waste Treatment Process	Main Characteristics and Assumptions
Incineration with energy recovery	Design capacity of 253,200 tons per year, where all the energy recovered will be used for the operation of the EIP and will produce 10% ash that will be landfilled.
Sorting and material recovery facility	Two lines of sorting with an annual capacity of 500,000 tons each. It is assumed that all the solid waste received is mixed and not being subjected for sorting at source. Collection and transportation to the EIP are excluded from the LCA analysis.
Composting	All the biodegradable organic fraction is separated and subjected to composting. The amount of the solid waste that will be subjected to composting is 453,250 tons per year (25% of the generated waste) to produce 300,000 tons per year of finished compost. All the produced compost is assumed to be used as a final cover of the landfill that is located within the EIP boundaries.
Landfilling	It is assumed that all the residual waste, after separation and energy recovery, will be directed to the sanitary landfill. The estimated landfilled waste is about 101,824 tons/year and assumed to be inert; it will not be subjected to biodegradation.

2.2.4. Impact Assessment

In the LCA impact assessment phase, the impact categories are identified, and their magnitude is assessed. In this study, impacts were modeled by the widely used midpoint model for the reduction and assessment of chemical and other environmental impacts, TRACI 2.1, which has been expanded and developed for sustainability metrics and expresses impacts in terms of discrete environmental effects [48]. This method covers the following impacts categories followed by their units for each:

- Global warming (kg CO_2 eq);
- Ozone depletion (kg CFC-11 eq);
- Smog formation (kg O_3 eq);
- Respiratory effects (kg PM2.5 eq);
- Acidification (kg SO_2 eq);
- Eutrophication (kg N eq);
- Toxic carcinogenic and noncarcinogenic substances (CTUh);
- Fossil fuel depletion (MJ surplus);
- Ecotoxicity (CTUe).

2.2.5. Interpretation

The interpretation stage includes presentation and analysis of the results. The highest impacts based on two scenarios are presented. To check the reliability and robustness of the results, sensitivity analysis was carried out. This was performed by investigating how the variation in the inputs values will affect the outputs. In this study, the impact of variation in the sorted solid waste quantity on the values of the emissions was assessed.

Figure 5 shows a flow diagram of the methodology followed in conducting the LCA analysis.

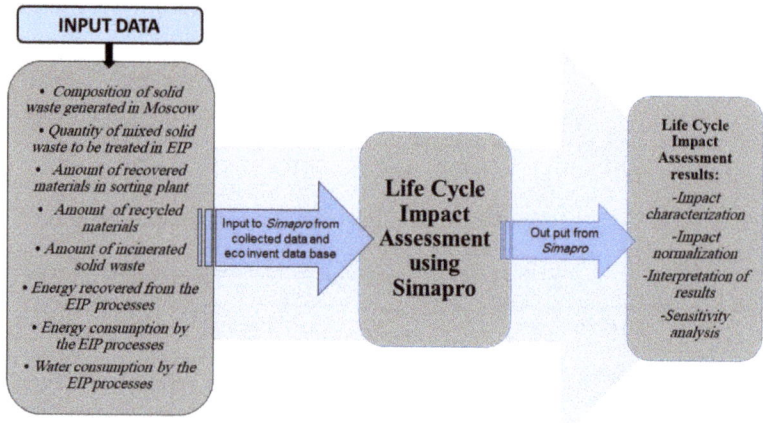

Figure 5. Flow diagram of the methodology followed in conducting LCA analysis.

3. Results and Discussion

3.1. Circular Economy in Russia

In Russia, the concept of circular economy is not mature enough, as is the case in many EU countries. The enabling drivers are not yet developed enough to move the country toward a more sustainable and efficient consumption and production system. The barriers and challenges that the adoption of CE in Russia faces come under three main categories as follows:

3.1.1. Administrative and Regulatory Barriers

Among the barriers that the adoption of a CE model in Russia faces are the absence of institutional support with appropriate structure and lack of public awareness and decision makers' knowledge [7,49,50]. Therefore, decision makers need to better understand the CE principles in order to develop an appropriate supportive framework.

Bogoviz et al. (2021) [51] proposed an organizational administrative model for a CE in Russia. According to the model, a structural hierarchy has been proposed with higher levels of decentralization and dependence, in which the state role is limited to issuance of regulations and standards, monitoring of compliance, and finance. One core principle of the CE is the efficient use of resources by the industry. According to a survey conducted by Ratner et al. 2021 [49], Russian firms are suffering from the complexity of administrative and regulatory procedures to increase resource efficiency, where the rules and regulations are outdated, which is leading to high cost of projects to adopt resource efficiency. As a result, the use of renewable energy by the Russian industrial firms is very limited. Only 3.8% of the companies covered by the survey conducted by Ratner et al., 2022 [49], are using renewable energy. According to Liubarskaia and Putinceva (2021) [52], Russia is lacking well developed legislation for secondary resources circulation. This suggests the urgent need for radical changes in the regulatory frameworks to create enabling conditions for the industry to improve their resource efficiency.

3.1.2. Knowledge and Awareness Barriers

One of the prerequisites to institutionalize sustainable solid waste management approaches based on the principles of circular economy is the availability of qualified and trained human resources. Educational institutions including universities are agents of change for sustainability [53], and play a vital role in overcoming the issues of lack of awareness, consumers' behavior, and knowledge in the field of CE [54,55]. In Russia, higher educational institutions can help in building capacity by preparing qualified human resources, in order to create a critical mass of experts that will enable the smooth transition of Russia towards a CE model. To achieve this objective, Russian institutions gradually started incorporating circular economy aspects into their curriculum (for example, RUDN University, Irkutsk National Research Technical University, Perm National Research Polytechnic University). In order to join the fragmented efforts, in 2021 the Russian environmental operator (REO), in collaboration with 15 Russian universities and leading companies specializing in solid waste management, formed a consortium for capacity building by training personnel in the field of CE. Moreover, within the framework of the working group on extended producer responsibility (EPR) representatives of leading higher education institutions, such as RUDN University and RANEPA, are actively involved in promoting the circular economy in training programs.

International agencies, such as the German International Association for Development (GIZ), also play a crucial role in the capacity building efforts. Within the framework of a project titled "Climate-Neutral Waste Management in the Russian Federation", GIZ Russia took the initiative to develop training materials for decision makers working in the Russian waste sector, to build their capacities in order to introduce the circular economy principles into the Russian waste management system.

3.1.3. Financial Barriers

Access to finance is a major issue for the enterprises engaged in CE to improve their sustainability performance. According to the International Financial Corporation of the World Bank [56], the main barrier to the development of waste industry in Russia is the lack of finance to construct waste processing facilities. Larchenko et al. (20121) [50] reported that economic barriers, including financial support to enterprises, to adopting CE are one of the serious obstacles. The smaller the size of the firm, the more difficulty there is in obtaining finance [57].

In the EU, extended producer responsibility (EPR) principle has emerged as a significant tool which fosters the enforcement of the circular economy package [58], while in Russia, EPR is one of the pillars of the new regulatory package of solid waste management. According to the Federal Law FZ 89 of the year 2014 on Waste Production and Consumption, the EPR was introduced as a financial mechanism to help in the transition towards a circular economy. However, the proposed structure of the EPR policy has certain shortcomings [9]. For example, the environmental fees under the proposed EPR system for PET and paper are far below the cost of obtaining materials ready for recycling, which hinders the financial sustainability of recycling such items. Another issue that faces the adoption of EPR is the absence of a well-established market for secondary materials. However, starting from 2018, several materials, such as scrap metals, paper, car tires, polymers, electrical appliances, and electronic waste, have been banned from disposal into landfills, yet the market for secondary materials is still in its initial development stage [40,43]. The need for the involvement of all stakeholders, including generators of solid waste, manufacturers, and regulatory agencies, is a major factor in the successful implementation of an EPR system. Creating economic incentives for the enterprises, especially medium and small (MSE) ones, to adopt CE initiatives will decrease the risks for such businesses in investing in recycling, recovery infrastructure, and eco-technologies for closing the loop.

3.2. Eco-Industrial Parks and Circular Economy

EIPs play a significant role in promoting circular economy models. In Russia, such projects help in the adoption of the CE concept to achieve sustainable development in the country [59]. According to Ratner et al. (2022) [49], one of the preferred options for Russian firms to promote the CE on a company level, is to have demonstration projects and to enhance cooperation with enterprises in other sectors for waste exchange and material reuse. EIP lends itself as an appropriate vehicle to demonstrate the collaboration of various industries residing at the EIP level. Therefore, developing and operating EIPs can serve to achieve the objectives of the circular economy. However, in the Russian Federation, attention should be directed not only for the development of EIPs, but also for the optimization of their operations to promote the circular economy model [40].

The exponential increase in the amounts of solid waste generation and the absence of infrastructure for source separation of municipal solid waste is a serious problem that is reflected in the efficient operation of EIPs in Russia [60], and consequently on the circular economy concept [59,61]. Realizing this fact, the Russian government decided to move from traditional end-of-pipe technology into a more sustainable and integrated solid waste management approach, where the circular economy is one of the core pillars [62]. To achieve that, a territorial waste management scheme was introduced to all major cities of the country, and regional operators were appointed to implement waste management activities including transportation, utilization, and final disposal. According to the new arrangements, a two-bin source separation system is being introduced in Moscow [39]. Such arrangement will pave the way for better performance of EIPs to achieve the circularity objectives.

3.3. Life Cycle Assessment Analysis

Life cycle assessment was carried out in this study to assess the environmental and health impacts of the case of EIPs in Russia based on two scenarios. The baseline scenario mainly represented the business-as-usual conditions, where the solid waste generated is being hauled to an unsanitary landfill which lacks leachate and biogas management systems, while scenario 2 is diverting an annual amount of 1,813,000 tons from the landfill to the EIP with different treatment and recovery options, as shown in Figure 3.

The results of LCA characterization analysis for each impact category for both scenarios are presented in Table 3. As can be seen from the characterization table, the landfilling scenario has higher impacts when compared with landfilling under EIPs in all environmental impact categories. This is in line with the findings of Rajcoomar and Ramejeawn, (2016) [31], who reported that landfilling scenarios have the highest values of impacts. As shown in Table 2, the total global warming potential of the landfilling scenario is 9.05×10^8 kg CO_2 eq. This global warming impact is mainly due to the fact that there is no gas control and management system in the landfill, where all the generated greenhouse gases are emitted to the atmosphere. Considering the total annual amount of the landfilled solid waste is 1.813 million tons, the per ton greenhouse gas emission is 500 kg CO_2 eq/ton. This is in agreement with the value that was reported by Vinitskaia et al. (2021) [39], who found that landfilling in Moscow region is emitting 0.5 t CO_2 eq per 1 ton of disposed solid waste. Another study by Abu Qdais et al. (2019) [63] found the greenhouse gas emissions from landfills is about 1115 kg CO_2 eq/ton, which is more than twice of the value reported in the current study. This is may be attributed to the fact that the study by Abu Qdais et al. (2019) [63] was conducted for Jordan, where the organic fraction of municipal solid waste (food) is greater than 50%, and the country is located in a hot arid climate, which is not the case in Moscow, where the organic percentage is 25% and the city is located in a cold region. In addition to the greenhouse gases emitted, there are other air pollutants emitted under both scenarios that include PM 2.5, chlorofluorocarbons (CFCs), and smog.

Table 3. Values of all impact categories for the two landfilling scenarios.

Impact Category	Unit	Landfilling (S1)	Landfill Eco-Industrial Park (S2)	Percent Reduction/Increase
Global warming	kg CO_2 eq	9.05×10^8	3.70×10^8	59%
Eutrophication	kg N eq	2,475,291	460,371.03	81%
Acidification	kg SO_2 eq	249,148.17	184,625.32	26%
Smog	kg O_3 eq	3,920,057.9	3,227,569.2	18%
Ozone Depletion	kg CFC-11 eq	6.9706184	1.2806006	81%
Carcinogenic	CTUh	3.7935234	0.26318803	92%
Non carcinogenic	CTUh	7.3697049	0.2988090	96%
Ecotoxicity	CTUe	1.64×10^8	5,372,424.5	96%
Fossil fuel depletion	MJ surplus	60,254,697	52,659,851	13%
Respiratory effects	kg PM2.5 eq	26,690.386	40,035.7	+33%

Under the second scenario, the landfill has a lower global warming potential than under the first scenario, as the biodegradable organic fraction of the solid waste is diverted from the landfill to be subjected for composting, where the landfill under this scenario receives only inert waste from the residues of the EIP processes with compost as a landfill cover. Moreover, it can be noticed from Table 3 that the second scenario (EIP) has less environmental and public health impacts by different percentages, except for the respiratory effect where the landfill scenario impact is less than the EIP scenario by 33%. The total global warming potential of the second scenario is less by 59% of the first scenario, while the eutrophication, acidification, smog, ozone depletion, carcinogenic, non-carcinogenic, eco-toxicity, and fossil fuel depletion impacts under the second scenario are reduced by 81%, 26%, 18%, 81%, 92%, 96%, 96%, and 13%, respectively.

Each process of the EIP has a contribution to the various impacts categories. Figure 6 presents the results of LCA characterization of the EIP processes, which shows the share of each EIP process to various impact categories. It can be seen that the landfill has the highest share in the eutrophication and global warming impacts with 100% and 95%, respectively. These findings are also confirmed by other researchers [30,34]. Under the EIP scenario, the landfilling contribution to impacts other than global warming is minimal, ranging from zero to 10%. On the other hand, it can be observed that solid waste sorting has an environmental benefit for all impact categories, where it ranges from 20% for eutrophication to 95% for the respiratory effect impact category. Organics composting has a minimal share in all impact categories, ranging from 2 to 5%.

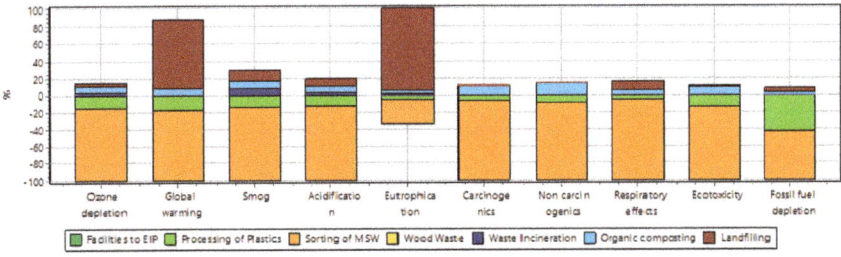

Figure 6. Contribution of each EIP process to various impacts categories.

Table 4 shows sensitivity analysis as a result of diverting 10% and 20% of the solid waste from the landfill to the sorting process at the EIP. It can be observed that the environmental impacts have been decreased by different percentages ranging from 22% to

698%. For example, diverting 10% of landfilled solid waste to a sorting facility at the EIP led to a reduction in the global warming impacts from 2.60×10^7 to -9.09×10^7 kg CO2 eq (i.e., 249% decreasing). This may be attributed to the decrease in the volume of raw material mining as a result of recycling metals, glass, and different types of plastics. These percentages were doubled when the 20% of landfilling was reduced. This explains the large difference in global warming impacts when the quantity of sorting increases. Vintiskaya et al. (2021) [39] reported emission reductions of 90–1850% during sensitivity analyses for municipal solid waste management systems in Moscow.

Table 4. Sensitivity analysis based on 10% and 20% diversion of solid waste from the landfill.

Impact Category	Unit	Emissions from EIP (Scenario 2)	Emission with 10% Conversion from Landfill	Emission with 20% Conversion from Landfill	Decrease in the Impact Category for 10% Diversion	Decrease in the Impact Category for 20% Diversion
Ozone depletion	kg CFC-11 eq	−16.721524	−21.146866	−25.572208	26.50%	53%
Global warming	kg CO_2 eq	26,045,152	−90,906,690	-2.08×10^8	249%	698%
Smog	kg O_3 eq	−14,728,371	−19,733,735	−24,739,099	33%	68%
Acidification	kg SO_2 eq	−1,486,174.7	−1,909,028.5	−2,331,882.3	28.50%	57%
Eutrophication	kg N eq	340,068.98	263,239.19	186,409.4	22.60%	45%
Carcinogenic	CTUh	−14.434815	−17.961323	−21.487831	24%	49%
Non carcinogenic	CTUh	−44.415678	−55.6601	−66.904522	25%	50.6%
Respiratory effects	kg PM2.5 eq	−291,183.18	−368,841.4	−446,499.62	26.7%	53%
Ecotoxicity	CTUe	-2.52×10^8	-3.13×10^8	-3.74×10^8	24%	49%
Fossil fuel depletion	MJ surplus	-5.15×10^8	-6.46×10^8	-7.77×10^8	25%	51%

4. Conclusions and Recommendations

Although circular economy models have been successfully adopted in many countries, the concept is still immature in the Russian Federation. The solid waste management in the country, until recently, was a disposal-driven system, where about 90% of the generated solid waste found its way in most cases to unsanitary landfill sites. Such practices are neither safe nor economic. The Russian government realized that such an approach does not serve the objectives of sustainability, and started making radical changes to the solid waste regulatory framework. The national solid waste management strategy calls for diverting 80% of the generated solid waste to recycling facilities by the year 2030, which will enhance the adoption of CE in the country. Despite this, there are still several challenges which need to overcome in order to secure a smooth transition to CE. Among such challenges is the low level of solid waste separation at source. Furthermore, the lack of enabling institutional framework and availability of human resources that are capable of leading the transition to CE are other issues that need to be resolved.

One of the aspects that could help in achieving sustainable solid waste management in Russia is the establishment of EIPs. Such enterprises divert significant amounts of items that exist in the solid waste stream from landfills to be reused and recycled as a result of material and energy recovery. By applying the LCA model to one of the operating EIPs in Russia, the study showed a decrease in both environmental and health impacts of an EIP scenario as compared to the baseline scenario where the waste disposed into landfills. Normalization of the assessed impacts has shown that eutrophication, carcinogenic, and global warming are the highest among all the impacts, while the ozone depletion, acidification, fossil fuel depletion, and smog impact categories are minimal.

The results of the study indicate the necessity of creating an enabling environment to efficiently adopt the CE principles in Russia. Russian universities are in a good position to lead the process of capacity building in institutions that are involved in the implementation of CE strategies. Such capacity-building efforts should be performed in a better collaboration with industry and public authorities. LCA is an effective tool to analyze waste management options based on their environmental performance. Applying the LCA analysis to one of the EIPS in Russia has shown that EIP recycling and recovery activities can reduce both environmental and health impacts, as compared to traditional waste management that relies mainly on landfilling. One of the limitations of the study is that the marketing of the products is not within the boundaries of the study. This is due to the absence of data on the marketing stage. Therefore, it is recommended that further LCA studies should be directed to assess EIPs' performance that includes marketing.

Author Contributions: Conceptualization, H.A.A.-Q. and A.I.K.; Data curation, A.I.K.; Investigation, H.A.A.-Q.; Methodology, H.A.A.-Q.; Project administration, H.A.A.-Q.; Software, H.A.A.-Q.; Supervision, H.A.A.-Q.; Visualization, A.I.K.; Writing—original draft, H.A.A.-Q. and A.I.K. All authors have read and agreed to the published version of the manuscript.

Funding: This research received no external funding.

Institutional Review Board Statement: Not applicable.

Informed Consent Statement: Not applicable.

Data Availability Statement: Not applicable.

Acknowledgments: This paper has been supported by the RUDN University Strategic Academic Leadership Program.

Conflicts of Interest: The authors declare no conflict of interest.

References

1. Korhonen, J.; Honkasalo, A.; Seppälä, J. Circular Economy: The Concept and its Limitations. *Ecol. Econ.* **2018**, *143*, 37–46. [CrossRef]
2. Ghosh, S.K.; Agamuthu, P. Circular economy: The way forward. *Waste Manag. Res.* **2018**, *36*, 481–482. [CrossRef] [PubMed]
3. Škrinjarí, T. Empirical assessment of the circular economy of selected European countries. *J. Clean. Prod.* **2020**, *255*, 120246. [CrossRef]
4. Rodriguez-Anton, J.M.; Rubio-Andrada, L.; Celemín-Pedroche, M.S.; Alonso-Almeida, M.D.M. Analysis of the relations between circular economy and sustainable development goals. *Int. J. Sustain. Dev. World Ecol.* **2019**, *26*, 708–720. [CrossRef]
5. Bogoviz, A.V.; Bruno, S. Will the Circular Economy Be the Future of Russia's Growth Model? In *Exploring the Future of Russia's Economy and Markets*; Emerald Publishing Limited: Bingley, UK, 2018; pp. 125–141.
6. Plastinina, I.; Teslyuk, L.; Dukmasova, N.; Pikalova, E. Implementation of circular economy principles in regional solid municipal waste management: The case of Sverdlovskaya Oblast (Russian Federation). *Resources* **2019**, *8*, 90. [CrossRef]
7. Gadzhiev, N.G.; Murzak, N.A.; Mitenkova, A.E.; Skripkina, O.V.; Konovalenko, S.A. Problems of development of the circular economy as a factor in Russia's sustainable development. *South Russ. Ecol. Dev.* **2020**, *15*, 155–164. [CrossRef]
8. Kalchenko, O.; Evseeva, S.; Evseeva, O.; Plis, K. Circular economy for the energy transition in Saint Petersburg, Russia. *E3S Web Conf.* **2019**, *110*, 02030. [CrossRef]
9. Wiesmeth, H.; Starodubets, N.V. The management of municipal solid waste in compliance with circular economy criteria: The case of Russia 1. *Econ. Reg.* **2020**, *16*, 725–738. [CrossRef]
10. Ilyina, L.A.; Garanina, M.P.; Ilyina, T.A.; Maslova, O.P. Creation of the Circular Economy in Russia as a Means of Acceleration Transition to the Market Path of Development. In *Circular Economy in Developed and Developing Countries: Perspective, Methods and Examples*; Emerald Publishing Limited: Bingley, UK, 2020; pp. 157–164. [CrossRef]
11. Kalioujny, B.; Ermushko, J.; Zhavoronok, A. Establishment of a strategy of circular economy increasing the well-being of society: Comparison of two national policies. *SHS Web Conf.* **2016**, *28*, 01050. [CrossRef]
12. Dorokhina, E.Y. Industrial and eco-industrial parks as a means for the resolution of regional conflicts in the use of natural resources. *Econ. Soc. Polit. Recreat. Geogr.* **2018**, *2*, 113–118. [CrossRef]
13. Li, W. Comprehensive evaluation research on circular economic performance of eco-industrial parks. *Energy Procedia* **2011**, *5*, 1682–1688. [CrossRef]
14. Martín Gómez, A.M.; Aguayo González, F.; Marcos Bárcena, M. Smart eco-industrial parks: A circular economy implementation based on industrial metabolism. *Resour. Conserv. Recycl.* **2018**, *135*, 58–69. [CrossRef]
15. Alabaeva, N.S.; Velitskaya, S.V.; Malahova, O.S.; Koroliova, C.P. Development of eco-industrial parks in Russia and abroad. *J. Econ. Bus.* **2019**, *6*, 19–22. [CrossRef]
16. Al-Quradaghi, S.; Zheng, Q.P.; Elkamel, A. Generalized framework for the design of eco-industrial parks: Case study of end-of-life vehicles. *Sustainability* **2020**, *12*, 6612. [CrossRef]
17. Chertow, M.R. Industrial ecology in a developing context. In *Sustainable Development and Environmental Management*; Springer: Dordrecht, The Netherlands, 2008; pp. 335–349. [CrossRef]

18. Mousqué, F.; Boix, M.; Négny, S.; Montastruc, L.; Genty, L.; Domenech, S. Optimal On-Grid Hybrid Power System for Eco-Industrial Parks Planning and Influence of Geographical Position. In Proceedings of the 28th European Symposium on Computer Aided Process Engineering, Graz, Austria, 10–13 June 2018; Elsevier Masson SAS: Issy-les-Moulineaux, France, 2018; Volume 43, ISBN 9780444642356.
19. Park, H.; Shah, I.H.; Gideon, T.N.; Lee, D.; Huong, T.T.; Park, Y.; Urjinkham, R. Eco-industrial park—A eco-innovation tool of circular economy transition in Korea. In Proceedings of the 2018 Circular Economy for Agro-Food Management, Ulsan, Korea, 13–15 June 2018.
20. Pomponi, F.; Moncaster, A. Circular economy for the built environment: A research framework. *J. Clean. Prod.* **2017**, *143*, 710–718. [CrossRef]
21. Fan, Y.; Fang, C. Assessing environmental performance of eco-industrial development in industrial parks. *Waste Manag.* **2020**, *107*, 219–226. [CrossRef]
22. Di Baldassarre, G.; Sivapalan, M.; Rusca, M.; Cudennec, C.; Garcia, M.; Kreibich, H.; Konar, M.; Mondino, E.; Mård, J.; Pande, S.; et al. Sociohydrology: Scientific Challenges in Addressing the Sustainable Development Goals. *Water Resour. Res.* **2019**, *55*, 6327–6355. [CrossRef]
23. Bellantuono, N.; Carbonara, N.; Pontrandolfo, P. The organization of eco-industrial parks and their sustainable practices. *J. Clean. Prod.* **2017**, *161*, 362–375. [CrossRef]
24. Zhao, H.; Zhao, H.; Guo, S. Evaluating the comprehensive benefit of eco-industrial parks by employing multi-criteria decision making approach for circular economy. *J. Clean. Prod.* **2017**, *142*, 2262–2276. [CrossRef]
25. Tian, J.; Liu, W.; Lai, B.; Li, X.; Chen, L. Study of the performance of eco-industrial park development in China. *J. Clean. Prod.* **2014**, *64*, 486–494. [CrossRef]
26. Wang, S.; Lu, C.; Gao, Y.; Wang, K.; Zhang, R. Life cycle assessment of reduction of environmental impacts via industrial symbiosis in an energy-intensive industrial park in China. *J. Clean. Prod.* **2019**, *241*, 118358. [CrossRef]
27. Valenzuela-Venegas, G.; Salgado, J.C.; Díaz-Alvarado, F.A. Sustainability indicators for the assessment of eco-industrial parks: Classification and criteria for selection. *J. Clean. Prod.* **2016**, *133*, 99–116. [CrossRef]
28. Azapagic, A.; Perdan, S. Indicators of sustainable development for industry: A General Framework. *Process Saf. Environ. Prot.* **2000**, *78*, 243–261. [CrossRef]
29. Belaud, J.P.; Adoue, C.; Vialle, C.; Chorro, A.; Sablayrolles, C. A circular economy and industrial ecology toolbox for developing an eco-industrial park: Perspectives from French policy. *Clean Technol. Environ. Policy* **2019**, *21*, 967–985. [CrossRef]
30. Erses Yay, A.S. Application of life cycle assessment (LCA) for municipal solid waste management: A case study of Sakarya. *J. Clean. Prod.* **2015**, *94*, 284–293. [CrossRef]
31. Rajcoomar, A.; Ramjeawon, T. Life cycle assessment of municipal solid waste management scenarios on the small island of Mauritius. *Waste Manag. Res.* **2017**, *35*, 313–324. [CrossRef]
32. Behrooznia, L.; Sharifi, M.; Hosseinzadeh-Bandbafha, H. Comparative life cycle environmental impacts of two scenarios for managing an organic fraction of municipal solid waste in Rasht-Iran. *J. Clean. Prod.* **2020**, *268*, 122217. [CrossRef]
33. Alamu, S.O.; Wemida, A.; Tsegaye, T.; Oguntimein, G. Sustainability assessment of municipal solid waste in Baltimore USA. *Sustainability* **2021**, *13*, 1915. [CrossRef]
34. Özer, B.; Yay, A.S.E. Comparative life cycle analysis of municipal waste management systems: Kırklareli/Turkey case study. *Environ. Sci. Pollut. Res.* **2021**, *28*, 63867–63877. [CrossRef]
35. Khandelwal, H.; Dhar, H.; Thalla, A.K.; Kumar, S. Application of life cycle assessment in municipal solid waste management: A worldwide critical review. *J. Clean. Prod.* **2019**, *209*, 630–654. [CrossRef]
36. Tulokhonova, A.; Ulanova, O. Assessment of municipal solid waste management scenarios in Irkutsk (Russia) using a life cycle assessment-integrated waste management model. *Waste Manag. Res.* **2013**, *31*, 475–484. [CrossRef] [PubMed]
37. Starostina, V.; Damgaard, A.; Rechberger, H.; Christensen, T.H. Waste management in the Irkutsk Region, Siberia, Russia: Environmental assessment of current practice focusing on landfilling. *Waste Manag. Res.* **2014**, *32*, 389–396. [CrossRef] [PubMed]
38. Kaazke, J.; Meneses, M.; Wilke, B.M.; Rotter, V.S. Environmental evaluation of waste treatment scenarios for the towns Khanty-Mansiysk and Surgut, Russia. *Waste Manag. Res.* **2013**, *31*, 315–326. [CrossRef] [PubMed]
39. Vinitskaia, N.; Zaikova, A.; Deviatkin, I.; Bachina, O.; Horttanainen, M. Life cycle assessment of the existing and proposed municipal solid waste management system in Moscow, Russia. *J. Clean. Prod.* **2021**, *328*, 129407. [CrossRef]
40. Yaroslavtsev, D.A.; Chekalin, V.S.; Liubarskaia, M.A. The Role of Information and Automation in Enhancing the Efficiency of the Functioning of Eco-industrial Parks in Russia. *Adv. Econ. Bus. Manag. Res.* **2019**, *47*, 933–937. [CrossRef]
41. Smirnova, T. And again defining the concept of Eco-Techno Park. *Waste Handl.* **2020**, *2*, 30–32.
42. Korotkovskaya, E.S.; Malinsky, I.G. Russian eco-technology parks as innovation element of the waste management system. In *Economic Development in the 21st Century, Proceedings of the II International Scientific Conference, Minsk, Belarus, 28 February 2020*; Belarus State University: Minsk, Belarus, 2020; pp. 165–167.
43. Putinceva, N.; Kim, O.; Voronina, E.; Fugalevich, E.; Mikhailova, M.; Ushakova, E. Introduction of innovative technologies—A factor in the development of the waste management industry in Russia. *Proc. IOP Conf. Ser. Mater. Sci. Eng.* **2020**, *940*, 012024. [CrossRef]
44. *ISO 14040:2006*; Environmental Management—Life Cycle Assessment—Principles and Framework. International Organization for Standardization: Geneva, Switzerland, 2006.

45. SimaPro SimaPro. Available online: http://www.pre-sustainability.com/simapro (accessed on 15 December 2019).
46. Nordelöf, A.; Poulikidou, S.; Chordia, M.; Bitencourt de Oliveira, F.; Tivander, J.; Arvidsson, R. Methodological Approaches to End-Of-Life Modelling in Life Cycle Assessments of Lithium-Ion Batteries. *Batteries* **2019**, *5*, 51. [CrossRef]
47. Bare, J. TRACI 2.0: The tool for the reduction and assessment of chemical and other environmental impacts 2.0. *Clean Technol. Environ. Policy* **2011**, *13*, 687–696. [CrossRef]
48. Moscow Register of MSW. *Territorial Scheme of Solid Waste Management in the City of Moscow*; Department of Housing and Communal Services of Moscow: Moscow, Russia, 2019.
49. Ratner, S.; Lazanyuk, I.; Revinova, S.; Gomonov, K. Barriers of consumer behavior for the development of the circular economy: Empirical evidence from Russia. *Appl. Sci.* **2021**, *11*, 46. [CrossRef]
50. Larchenko, L.; Paranina, A.; Kulachinskaya, A.; Kuramshina, L. Opportunities and Challenges of Transitioning to a Circular Economy Model in Russian Regions. In *Proceedings of the ICEPP 2021, Efficient Production and Processing*; Vankov, Y., Ed.; Springer Nature: Cham, Switzerland, 2022; Volume 190, pp. 291–297.
51. Bogoviz, A.V.; Revzon, O.A.; Poliakova, V.V.; Sumbatyan, S.L.; Morozova, N.G. Successful Manifestations of the Circular Economy in Modern Russia. In *Circular Economy in Developed and Developing Countries: Perspective, Methods and Examples*; Emerald Publishing Limited: Bingley, UK, 2021; pp. 149–156. [CrossRef]
52. Liubarskaia, M.A.; Putinceva, N.A. The Role of Secondary Resource Market in the Development of the Extended Producer Responsibility Mechanism in Russia. *IOP Conf. Ser. Earth Environ. Sci.* **2021**, *938*, 012014. [CrossRef]
53. Abu Qdais, H.; Saadeh, O.; Al-Widyan, M.; Al-tal, R.; Abu-Dalo, M. Environmental sustainability features in large university campuses: Jordan University of Science and Technology (JUST) as a model of green university. *Int. J. Sustain. High. Educ.* **2019**, *20*, 214–228. [CrossRef]
54. Qu, D.; Shevchenko, T.; Saidani, M.; Xia, Y.; Ladyka, Y. Transition towards a circular economy: The role of university assets in the implementation of a new model. *Detritus* **2021**, *17*, 3–14. [CrossRef]
55. Salas, D.A.; Criollo, P.; Ramirez, A.D. The Role of Higher Education Institutions in the Implementation of Circular Economy in Latin America. *Sustainability* **2021**, *13*, 9805. [CrossRef]
56. IFC. Waste in Russia: Garbage or Valuable Resource? Scenarios for Developing the Municipal Solid Waste Management Sector, International Finance Corporation, World Bank. 2014. Available online: https://documents1.worldbank.org/curated/pt/702251549554831489/pdf/Waste-in-Russia-Garbage-or-Valuable-Resource.pdf (accessed on 24 January 2022).
57. Aranda-Usón, A.; Portillo-Tarragona, P.; María Marín, L.; Scarpellini, S. Financial Resources for the Circular Economy: A Perspective from Businesses. *Sustainability* **2019**, *11*, 888. [CrossRef]
58. Pouikil, K. Concretising the role of extended producer responsibility in European Union waste law and policy through the lens of the circular economy. *ERA Forum* **2020**, *20*, 491–508. [CrossRef]
59. Saha, I.; Tatiana, S.; Smirnova, T.S.; Vladimir, A.; Maryev, V.A. Implementation of Eco-Industrial Park for Effectual Establishment of Circular Economy in Russia. *Nat. Environ. Pollut. Technol.* **2021**, *20*, 2031–2040. [CrossRef]
60. Vorotnikov, A.M.; Bajanov, I.N.; Lijin, D.H. Eco-Techno Parks: Development Prospects, Priority areas of implementation and features of Financing. *Probl. Natioal Strategy.* **2019**, *55*, 144–155.
61. Bobylev, S.N.; Solovyeva, S.V. Circular economy and its indicators for Russia. *Mir Novoi Ekon. = World New Econ.* **2020**, *14*, 63–72. [CrossRef]
62. Kurbatova, A.; Abu-Qdais, H.A. Using multi-criteria decision analysis to select waste to energy technology for a Mega city: The case of Moscow. *Sustainability* **2020**, *12*, 9828. [CrossRef]
63. Abu Qdais, H.A.; Wuensh, C.; Dornack, C.; Nassour, A. The role of solid waste composting in mitigating climate change in Jordan. *Waste Manag. Res.* **2019**, *37*, 833–842. [CrossRef] [PubMed]

Article

Solid Waste Management in the Context of a Circular Economy in the MENA Region

Safwat Hemidat [1,*], Ouafa Achouri [2], Loubna El Fels [3], Sherien Elagroudy [4], Mohamed Hafidi [3], Benabbas Chaouki [5], Mostafa Ahmed [4], Isla Hodgkinson [6] and Jinyang Guo [7]

1. Department of Waste and Resource Management, Rostock University, 18051 Rostock, Germany
2. Department of Environmental Engineering, Faculty of Process Engineering, University of Salah Boubnider Constantine 3, 'B' 72 Ali Mendjeli Nouvelle Ville, Constantine 25000, Algeria; ouafa.achouri@univ-constantine3.dz
3. Laboratory of Microbial Biotechnologies, Agrosciences and Environment (BioMAgE) Labeled Research Unit-CNRST N°4, Cadi Ayyad University, Marrakesh 40000, Morocco; loubna.elfels@gmail.com (L.E.F.); hafidi@uca.ac.ma (M.H.)
4. Egypt Solid Waste Management Center of Excellence, Faculty of Engineering, Ain Shams University, Abbasseya, Cairo Governorate 11535, Egypt; s.elagroudy@eng.asu.edu.eg (S.E.); mostafa.sattar@gmail.com (M.A.)
5. Institute of Urban Techniques Management, Constantine 3 University, 'B' 72 Ali Mendjeli Nouvelle Ville, Constantine 25000, Algeria; benabbas.chaouki@gmail.com
6. Institute of Waste Management and Circular Economy, Technical University of Dresden, 01069 Dresden, Germany; isla_marie.hodgkinson@tu-dresden.de
7. Sustainable Resource and Waste Management, Hamburg University of Technology, Blohmstr. 15, 21079 Hamburg, Germany; jy.guo@tuhh.de
* Correspondence: safwat.hemidat2@uni-rostock.de

Citation: Hemidat, S.; Achouri, O.; El Fels, L.; Elagroudy, S.; Hafidi, M.; Chaouki, B.; Ahmed, M.; Hodgkinson, I.; Guo, J. Solid Waste Management in the Context of a Circular Economy in the MENA Region. *Sustainability* 2022, 14, 480. https://doi.org/10.3390/su14010480

Academic Editors: Hani Abu-Qdais, Anna Kurbatova and Caterina Picuno

Received: 23 November 2021
Accepted: 30 December 2021
Published: 3 January 2022

Publisher's Note: MDPI stays neutral with regard to jurisdictional claims in published maps and institutional affiliations.

Copyright: © 2022 by the authors. Licensee MDPI, Basel, Switzerland. This article is an open access article distributed under the terms and conditions of the Creative Commons Attribution (CC BY) license (https://creativecommons.org/licenses/by/4.0/).

Abstract: Solid waste management in most MENA countries is characterized by lack of planning, improper disposal, inadequate collection services, inappropriate technologies that suit the local conditions and technical requirements, and insufficient funding. Therefore, waste management is mainly limited to collection, transportation, and disposal. As the circular economy has recently been given high priority on the MENA region's political agenda, all MENA member states are seeking to move away from old-fashioned waste disposal, "waste management", towards a more intelligent waste treatment, "resource efficiency". This paper presents a comprehensive overview of national systems for municipal solid waste (MSW) management, and material and energy recovery as an important aspect thereof, in the context of the circular economy in selected countries in the MENA region. Since policy, regulation, and treatment technologies are traditionally connected to MSW management, the focus of this article is twofold. Firstly, it aims to identify the different practices of solid waste management employed in selected MENA region countries and their approaches to embracing the circular economy and, secondly, it examines the extent to which policies and technologies applied play any role in this context. The study revealed that most waste management issues in the countries analyzed appear to be due to political factors and the decentralized nature of waste management with multi-level management and responsibilities. In fact, material and energy recovery in the context of municipal solid waste management does not differ significantly in the countries in the MENA region considered. In most cases, "waste" is still seen as "trouble" rather than a resource. Therefore, a fresh vision on how the solid waste management system can be transformed into a circular economy is required; there is a need for paradigm shift from a linear economy model to a circular-economy model.

Keywords: solid waste management (SWM); legal framework; financial framework; treatment technologies and disposal; MENA region

1. Introduction

The MENA region covers an area of more than 15 million square kilometers and has about six percent of the world's population, equivalent to the population of the European Union (EU). The total population of the region has increased from about 100 million in 1950 to about 465 million in 2020 [1,2].

The MENA region is characterized by its dependence on its non-renewable resources It includes 20 countries, with a total population of about 465 million (Table 1). The three smallest countries, Bahrain, Djibouti, and Qatar, each have populations of 1.7 million, 1.0, and 2.9 million, respectively. In contrast, the two largest countries, Egypt and the Islamic Republic of Iran, have about 102 and 84 million people, respectively. Along with Algeria, Morocco, and Sudan, these five most populous countries account for about 70% of the region's population. About half of the population lives in cities [1].

Table 1. Population of MENA region countries [1].

MENA Member Country	Population in Million Inhabitants	MENA Member Country	Population in Million Inhabitants
Egypt	102	UAE	9.8
Iran	84	Israel	9
Algeria	44	Lebanon	7
Iraq	40	Libya	7
Morocco	37	Oman	5
Saudi Arabia	35	West Bank & Gaza	4.8
Yemen	30	Kuwait	4
Syria	17.5	Qatar	2.9
Tunisia	12	Bahrain	1.7
Jordan	10.2	Djibouti	1

Currently, solid waste management (SWM) is one of the most important environmental challenges facing countries in the Middle East and North Africa (MENA) region. This is attributed to the rapid growth of the population, a booming economy, rapid urbanization, and high standards of living in the community, which have significantly accelerated the rate of solid waste generation [3,4]. The provision of an efficient and sustainable waste management system that takes into account the potential impact on public health and the environment is critical to most governments in the region.

In general, the waste resource sector across the region is insufficiently regulated. Notwithstanding continuous attempts, most countries in the MENA region have not yet put in place appropriate long-term legislation and strategies in the field of waste management and the circular economy [4]. The waste management system in the region faces many obstacles, for example, but not limited to, lack of planning, centralization of authority at the national level, overlapping of powers and responsibilities, lack of coordination between relevant institutions and ministries, insufficient cost recovery mechanisms, lack of trained and skilled personnel, inequality in service between rural and urban areas, and the lack of reliable databases [5,6]. The inability of the existing waste management systems and municipal administrations to cope with the growing waste-generation rates and the pressures placed on waste infrastructure has led to significant health and environmental problems in the region. Demand for municipal services continues to outstrip the capacity of systems and infrastructure that have already seen years or even decades of under-investment, unreliability, and high costs [7,8].

Despite the fact that, recently, many countries in the MENA region have introduced the concept of Integrated Solid Waste Management (ISWM), to date, there have been no tangible results in terms of reducing the amount of waste that is landfilled or dumped. The latter remains the primary waste disposal practice in most MENA countries, being poorly managed and lacking most basic engineering and sanitary procedures for collecting and treating gas and leachate. However, collection and sorting, composting, incineration of medical waste, and sanitary landfills are starting to be practiced, while recycling, reuse, and resource recovery are still in the initial stages [7–9]. Against this background, a fresh look is required and there is a need for a new a joint vision on how the solid waste management system can be transformed into a circular economy. Being the main drivers of the transition towards a sustainable waste management system, the policies, strategies, and practices in place that regulate the performance of waste management in the region should be comprehensively reviewed. To this end, the aim of this study was to identify the different practices of solid waste management employed in selected MENA region countries and their approaches to embracing the circular economy. A further objective was to examine the extent to which the applied waste management policies, regulations, and technologies play any role in the context of circular economy.

2. Municipal Solid Waste Generation and Composition in MENA Region Countries

The definition of municipal solid waste (MSW) can vary widely between countries. Typically, the term municipal solid waste refers to solid waste generated from community activities (for example, residential, commercial, institutional, and industrial). While construction waste and hazardous waste are excluded from MSW in European countries, they are considered to be domestic solid waste in most developing countries [10]. Municipal solid waste is commonly called waste, garbage, trash, or refuse; hence, waste refers to the waste generated from several activities.

The amount of MSW generated by developing countries is about 120 million tons per year. This amount was generated by 16 countries (Table 2). The average values of waste generation per capita in developing countries are relatively low, ranging from 0.5 to 1.1 kg/capita/day, compared to those countries in the Organization for Economic Co-operation and Development (OECD) with an average value of 2.2 kg per capita per day [4].

Table 2. Municipal solid waste amount in the MENA region.

Country	Waste kg/Day		Waste Volume	Disposal Route %	
	Large City	Rural	Million Ton/Year	Recovery	Landfilling
Egypt	0.85	0.5	26.0	10	90
Algeria	0.8	0.6	13.5	8	92
Bahrain	2.7	1.1	0.8	5	95
Iraq	1.4	0.85	12.8	5	95
Yemen	0.6	0.35	3.8	7	93
Jordan	0.9	0.6	3.4	7	93
Kuwait	1.5	1.4	1.6	10	90
Lebanon	0.95	0.8	2.04	23	77
Libya	1.12	0.85	3.2	3	97
Morocco	0.76	0.3	7.4	9	91
Qatar	2.5	0.6	1.5	10	90
Oman	1.3	0.7	2.5	5	95
Saudi Arabia	1.8	1.5	18	12	88
Syria	0.5	0.4	3.6	25	75
Tunisia	0.8	0.5	2.5	9	91
UAE	1.47	1.2	8.5	27	73

Having accurate and reliable data on waste composition is a critical factor for waste management and utilization options. Moreover, the process of selecting an approach for MSW management and treatment (recycling, energy recovery, etc.) is strongly influenced by the composition, characteristics, and composition of the waste. Thus far, these required data are not available in many developing countries, and if available they are in many cases inconsistent. Moreover, most of the available data are based on theoretical estimation rather than actual measurements [11,12].

A full description of the amount of waste generated and its composition in the four targeted MENA region countries is provided next.

2.1. Algeria

Given its surface area, Algeria is a large country; it is ranked tenth-largest in the world by size. However, it has a very varied population distribution, with high population densities in the north and very low in the south. The economic activities are concentrated in urban areas and large metropolises; consequently, there is a major difference in the waste generation between the cities in the country [13].

The total amount of municipal solid waste produced in the year 2020 is estimated to be around 13.5 million tons. The daily amount of waste generation per capital is about 0.80 kg/person/day. Municipal/household waste accounts for about 90% of all households and the remaining 10% corresponds to other waste streams. Indeed, the amount of waste generated is affected by several factors, the most important of which are the city's population and its nature (commercial, industrial, etc.) [14].

Availability of accurate and reliable data on the composition of the waste is essential and crucial in evaluating and selecting appropriate waste treatment options. Waste composition, density, and moisture are important factors for managing treatment facilities and also affect the waste collection process.

In this context, in 2014 and 2018, the Agence National des Déchets (National Waste Agency) conducted two major national campaigns to determine the composition of the waste generated in Algeria. The campaign, which was launched in April 2018, took a year (until the end of March 2019) to monitor the differences in the composition of the composition on a quarterly basis (Figure 1). It aimed to accurately determine the composition of household waste and obtain updated data that would enable determining the most appropriate waste treatment methods or technologies [14].

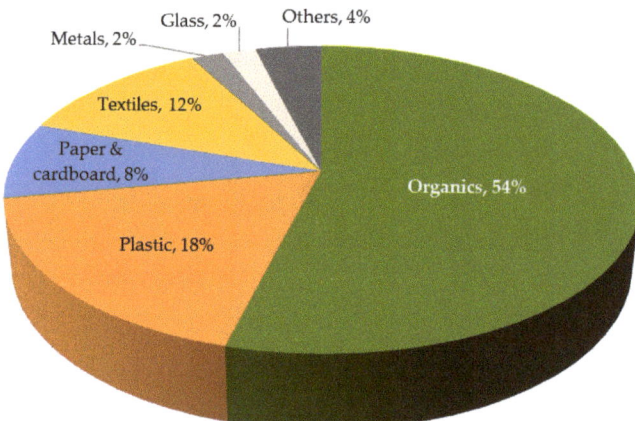

Figure 1. Average annual composition of MSW in Algeria [14].

The results show that organic materials dominate the waste composition, accounting for 54%, followed by plastics (18%), paper and cardboard (8%), textiles (12%), glass (2%), metals (2%), and other (4%). Recently, a transition has been made from the common disposal practices applied in Algeria (illegal dumping) towards controlled sanitary landfills, which reflects a real awareness of environmental protection and the need for an integrated management of municipal solid waste. Currently, the waste generated in Algeria is served through 122 to 146 technical landfills, 32 controlled landfills, 29 sorting centers, and 54 Class 3 landfill technical centers (for inert waste), in addition to the rehabilitation of 40 unauthorized dumpsites [14].

Regarding the recycling activities at the national level, compared to other sectors, ferrous metals constitute the most important recycling sector, where the amount recovered annually is about 628,915 tons. This includes scrap, steel, and iron castings.

As for the recycled plastic and paper and cardboard waste, the amounts are about 304,321 and 108,396 tons, respectively, annually. In addition, similar amounts of non-ferrous metals and wood are also recycled [14].

2.2. Egypt

Egypt generates around 26 million tons of MSW annually [15], thus representing almost one-third of the solid waste types generated in the country [16]. The MSW generation rates vary greatly between rural and urban areas. About 45% of the generated MSW comes from Greater Cairo (Cairo, Giza, and Qalioubiya governorates) and Alexandria, with a total population of 30.7 million generating 32,570 tons/day and with a collection efficiency between 50 and 70%. These regions are followed by the Delta region consisting of the seven governorates of Beheira, Kafr El-Sheikh, Gharbia, Monufia, Sharqia, Dakahlia, and Damietta. This area has a total population of 36.4 million, which generates 30% of the total MSW (Figure 2).

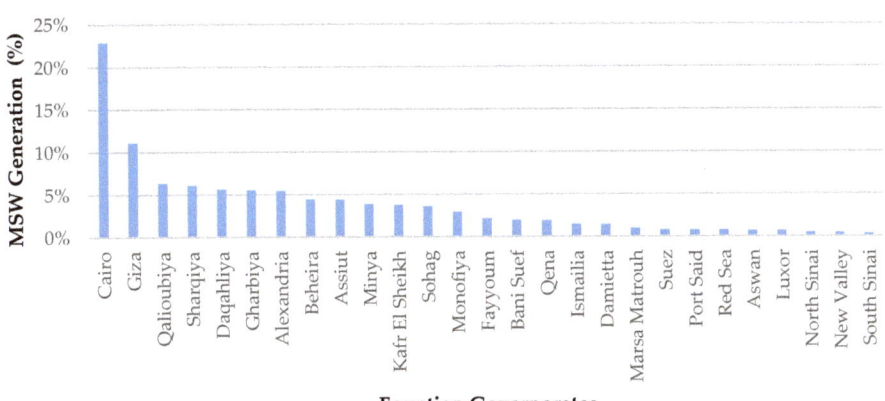

Figure 2. MSW generation distribution in Egypt [15].

The MSW collection coverage of the Delta region is only 50%, as the area is challenged by the shortage of public land for waste treatment due to the agricultural activities and land ownership, leading to a great deal of waste being dumped on vacant land or agricultural drains, causing severe negative environmental and public health implications. Lastly, the other 16 governorates, with a population of 35 million, generate the remaining MSW, which represents 25% of the waste generated. On these bases, MSW generation rates across the Egyptian governorates vary between 0.3 and 2.0 kg/capita/day [17,18] (CAPMAS, 2021; MoE, 2020).

The MSW composition in Egypt is 56% organics, 13% plastics, 10% paper and cardboard, 4% glass, 2% metals, and 15% other material [16], as shown in Figure 3. However, MSW composition varies significantly across the different governorates, where the organic fraction can range between 41 and 70%, plastic between 6 and 16%, glass between 1.5 and 9.4%, and metals between 1 and 8% [19].

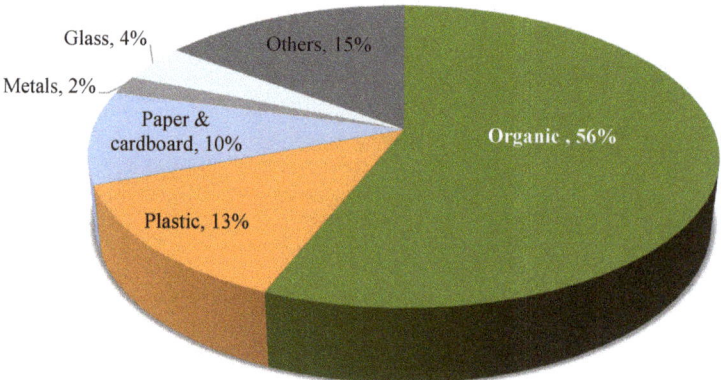

Figure 3. Average annual composition of MSW in Egypt.

Concerning the current waste management practices in Egypt, according to the most recent statistics issued by the Egyptian Ministry of Environment, MSW management in 2016 was based on open dumping (81%), a small recycling share (12%), and only 7% was landfilled [16]. Nevertheless, since its establishment in 2018, the Egyptian Waste Management Regulatory Authority (WMRA), has been working hard to improve its waste management status through enhancing recycling rates, by promoting composting, and RDF production. Currently, it is assumed that less than 10% of the OFMSW generated in Egypt is composted as the process is not economically feasible [20], while for RDF production, the Egyptian government is seeking to encourage production and its utilization in cement plants [21].

2.3. Jordan

The solid waste generation in Jordan has been continuously increasing. However, due to the influx of Syrian refugees, the country has witnessed a sharp increase in the amount of solid waste generated [8,22]. The amount of solid waste collected in 2015 (3,365,261 tons) is greater by 24% than that collected in 2013 [8,23].

As is the case in most developing countries, the solid waste generated in Jordan is mainly composed of organic materials. Figure 4 shows the composition of solid waste generated by different sources in the City of Irbid [22].

2.4. Morocco

Morocco generates 7.4 million tons of municipal solid waste annually. The solid waste generation per capita is estimated at 0.76 kg/day/capita and 0.3 kg/day/capita in urban and rural areas, respectively. About 70% of household waste produced in Morocco is organic, with a moisture content of 70%, which causes the leachate problem. The proportion of recyclable waste ranges between 15 and 25%, 5–10% of which is paper and cardboard, 6–10% plastic, 1–4% metal, and 1–2% glass (Figure 5).

The total amount of organic waste in Morocco is estimated at 68.6 to 90.04 million tons annually including food industry, agricultural residues, wood residues, and other organic waste. Sewage sludge is estimated to be about 234,000 tons in 2020. The amounts of industrial waste and demolition and construction waste are about 5.4 million tons and 14 million tons, respectively. As for medical waste, it is about 21,000 tons/year, of which 6000 tons are hazardous waste [24].

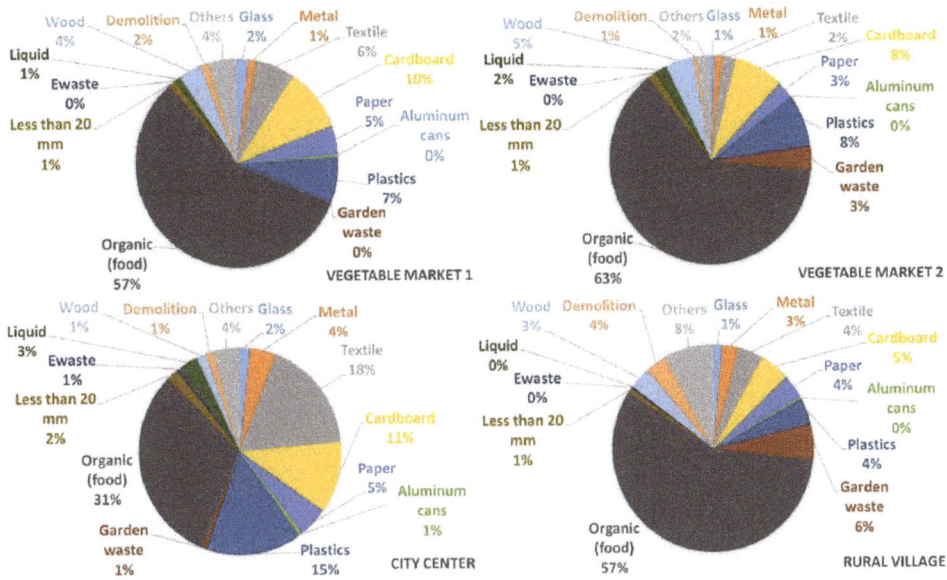

Figure 4. Solid waste composition generated by different sources in Irbid City, Jordan.

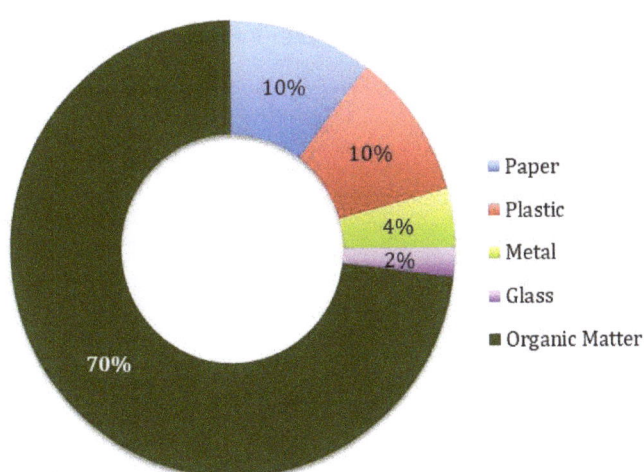

Figure 5. Solid waste composition generated by different sources in Morocco.

3. Municipal Solid Waste-Management Practices

Some waste management practices are more costly than others, and integrated approaches facilitate the identification and selection of low-cost solutions. Some waste management activities cannot bear any charges; some will always be net expenses, while others may produce an income. An integrated system can result in a range of practices that complement each other in this regard [25]. This means that the integrated solid waste management hierarchy cannot be followed strictly since, in particular situations, the cost of a prescribed activity may exceed the benefits, when all financial, social, and environmental considerations are taken into account. The MSW practices can be divided into four main activities:

- Sorting and collection: Waste sorting is the process of separating MSW into different types. Waste sorting can occur before or after the waste is collected. The collection process involves collecting waste from households, from community and street bins, or from bulk generators in large containers or vehicles. It extends to activities such as driving between stops, idling, loading, and on-vehicle compaction of waste.
- Recycling: After waste sorting, recyclables are reprocessed into products.
- Transfer and transportation: This process involves the delivery of collected waste to transfer stations or treatment facilities.
- Treatment and disposal: Waste treatment is the process of disposing of waste after collection. Waste can be buried at landfills or burned through an incineration process. Non-recyclable waste items can be converted into compost or energy as various forms of useable heat, electricity, or fuel.

These four activities are used to examine and analyze the current solid waste management practices in the four selected MENA region countries, as described next.

3.1. Algeria

Up to 97% of the waste generated in Algeria is send to different controlled landfills and dumping sites without any pre-treatment, which generates high levels of methane gas due to the high amount of organic and water content. Specifically, 57% of the generated waste is disposed in open landfills or dumpsites, and 30% is being burned in uncontrolled open public or municipal landfills, while only 10% is disposed of in controlled or sanitary landfills (Figure 6). On the other hand, the rate of recyclable material recovery from the waste stream does not exceed 3%; only 1% of the waste generated is recycled and 2% is subject to compost production [26]. It is clearly seen from Figure 6 that 87% of the generated waste is disposed of in open / uncontrolled dumpsites. According to a survey conducted by the Office of the Ministry of Regional Planning and Environment, more than 3130 surface dumps have been identified in the country, with a surface area of about 4552.5 hectares. Most of these dumps have almost reached their maximum capacity and cannot receive more waste [26].

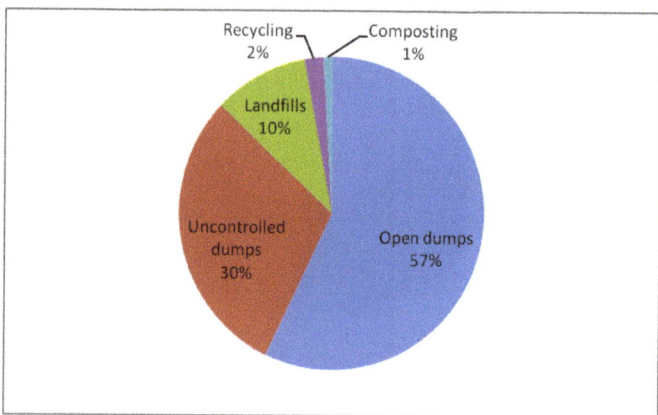

Figure 6. Methods of waste disposal in Algeria [27].

In 2001, the Algerian government decided to abandon the traditional model of waste disposal through open landfills and move towards the use of sanitary landfills. Accordingly, a plan was drawn up to establish 100 sanitary dumps, most of which were completed at the end of 2010 [27].

3.2. Egypt

Current waste management practices in Egypt do not include any source reduction or segregation programs. However, change in consumers' and residents' consumption patterns is often discussed as the cause of waste management challenges. These include increased waste-generation rates, increased use of single-use plastics, the difficulty of managing and treating mixed waste streams, inefficient use of resources, and, consequently, increased rates of pollution and environmental degradation [16,28,29]. However, recently, source reduction has been discussed extensively for managing single-use plastics, yet there have still been no source reduction measures implemented.

In terms of waste collection, the collection coverage was 20% in 2015 and reaching a maximum of 70% in Cairo city, the Egyptian capital. Improving waste collection is one of the main waste management plans in Egypt, which aims to reach 80% by the year 2030. These goals are also associated with the African Union Commission's Agenda 2063, which calls for improving urban waste collection services to cover nine out of ten people by 2023 [28,30]. Waste collection in urban areas in Egypt is conducted through an integration of formal and informal sectors, where waste is formally collected from waste containers installed in the streets for a fee collected from residential and commercial units, as discussed earlier [18]. However, the informal sector offers a door-to-door collection service in the neighborhoods where the residents can afford to pay an extra fee for such services; recyclables are picked out from the collected waste before disposing of it in the formal waste management chain. Generally, it can be assumed that the collection costs in Egypt represent between 50 and 80% of the MSW management costs.

According to the MSW collection system practiced in Egypt, most of the recycling activities are conducted through the informal sector and waste scavengers. The remaining waste in the formal waste management chain is commonly managed to produce compost and RDF, and to dispose of the rejects through landfilling and dumping. The OFMSW represents up to 70% of the MSW and its typical management approach is composting; however, that is challenged by the lack of appropriate compost quality that can comply with the environmental standards. Since there are no MSW source separation practices implemented in Egypt and MSW is collected as a mixed waste stream, there are substantial concerns about inadequately produced compost which comprises heavy metals that exceed the limitations of the Egyptian regulations [31,32]. Additionally, the process is not financially viable due to the compost's low sale value at 50–80 EGP per ton after production costs of up to EGP 200 [20]. Finally, for landfilling, Egypt is currently transitioning from disposing of waste in illegal dumpsites to controlled and sanitary landfills [15,33].

3.3. Jordan

The waste sector in Jordan is governed by several national authorities including municipalities, the Ministry of Local Administration (MOLA), and the Ministry of Environment (MoEnv).

Currently, Jordan generates around 3.2 million tons per year. On average, Jordanians produce 0.81 kg of municipal waste per capita per day. Up to 90% of the solid waste generated in Jordan (approximately 2.1 million tons) is disposed of at one of the existing 17 operating landfills. The Ghabawi landfill is the only landfill in Jordan that meets international best practices. It was developed after a feasibility study. It has been subjected to an environmental-impact assessment, and meets international design and construction standards [5,34]. Currently, Al-Akeeder, which is the second-largest landfill in the country and serves the northern Jordan, is being subjected to rehabilitation activities, where the old cells are closed, while the newly opened cells are meeting sanitary landfilling standards.

Municipalities are directly responsible for the daily collection of waste, in addition to bearing the full costs of the process of collection, transportation, and disposal in landfills. This usually consumes 70% of the municipal budget allocated to waste management. Thus, these duties can affect the limited financial resources available to cover these costs, distracting from thinking about the indirect costs.

The composition of municipal waste in Jordan is moving from a primarily organic mixture to a more complex mixture with more plastic, paper, and cardboard, as well as electronic waste. The composition of waste varies across the country, but in municipalities it is generally 51% organic, 15% plastic, and 14% paper. More than 50% of the waste generated by households is organic/food waste [9].

Recycling is still not practiced in Jordan on a formal level and the MSW is collected as mixed waste without any source separation activities. The solid waste recycling rate, which is 7%, is low, even when compared to the average of 10% in the Gulf Cooperation Council (GCC) countries. Other waste types, such as agricultural waste [35], hazardous waste, medical waste [36], construction and demolition waste, and electronic waste are also generated in increasing volumes in Jordan, with insufficient means of treatment and disposal. In terms of environmental impact, landfill waste is an important component of Jordan's greenhouse gas emissions profile, contributing 10% of the country's total greenhouse gas emissions [34].

Recently, the government has started moving towards improved practices of solid waste landfilling by gaseous-emission reduction and leachate containment by both natural and synthetic liners. The Ministry of Local Administration is responsible for overseeing the implementation of the National Strategy and Action Plan (NSAP) for Municipal Solid Waste Management 2015–2034, which prioritizes mitigating environmental degradation through rehabilitation of dumpsites/unsanitary landfills, isolating the landfills by appropriate fences and buffer zones, adding sanitary linings to manage the leachate, and capping to reduce greenhouse gases (GHG) emissions.

3.4. Morocco

Nowadays, Morocco generates 7.4 million tons of municipal solid waste annually, 79% of which is generated in urban areas. The per capita solid waste generation is estimated at 0.76 kg/day/capita and 0.3 kg/day/capita in urban areas and rural areas, respectively. However, the rate of solid waste production varies from city to city according to its nature and classification (residential, commercial, industrial, etc.). Taking into account the annual growth of municipal solid waste generation of about 1.36% and a collection rate of 96%, the total amount of solid waste landfilled is estimated to be around 26.8 million tons. According to figures published in the National Strategy for Waste Reduction and Recovery in 2019, the amount of landfilled waste is expected to reach 39 million tons by 2030. In 2017, the amount of household waste generated was about 4.7 million tons, in addition to 5.4 million tons of industrial waste and 14 million tons of construction and demolition waste [24].The amount of organic waste disposed in Morocco ranges between 68.6 and 90.04 million tons annually, including food industry, agricultural residues, wood residues, and other organic waste. The amount of industrial waste generated annually is about 5.4 million tons, in addition to 14 million tons of construction and demolition waste [24]. Organic waste dominates the waste composition, reaching about 70%, with a high water content of about 70%. The proportion of recyclable materials in the generated waste ranges between 15 and 25%, of which 5–10% is paper and cardboard, 6–10% plastic, 1–4% metal, and 1–2% glass. In terms of waste treatment and disposal, 25 disposal and recovery centers have been established since 2008. In the "Sorting, Recycling and Recovery" sector, priority is given to energy recovery from waste through anaerobic digestion, landfill gas production from sanitary landfills, incineration, and RDF utilization, while recycling is still in the initial stages [37]. Since the introduction of the National Program for Domestic Waste (PNDM) in 2008, Morocco has significantly improved the landfilling rate of domestic waste in controlled landfills, from 10% before 2008 to 44% in 2015 [38]. The current recycling rate is presented in Table 3. The remainder of the collected waste is deposited in the 300 illegal landfills in the country. Similarly, the rate of waste collection, which is operated by private companies under a public service license, has increased from 44% in 2007 to 86% in 2015. From a financial point of view, the collection costs in the municipal solid waste management budget are as royalty and range from 45 to 110 USD per ton, while landfilling costs range from 11 to 32 USD per ton [4,39]. Since the activation of the National Program for Domestic Waste (PNDM) in 2008, about 25 landfill and waste recovery (CEV) centers have been established across the country. A target of establishing 15 operational sorting centers in the CEVs by 2025, and 25 in 2030, has been set by the SNRVD. The first waste sorting center opened in September 2018 in Fez, ahead of the opening of the country's largest sorting center in Marrakesh. The production capacity of these facilities is between 300 and 768 tons of waste per day, which makes it possible to strengthen the sorting, recycling, and recovery channels required by the SNRVD and develop the capacity of local communities in terms of waste treatment [24,39].

Table 3. Current recycling rate (2015 data) and new SNRVD objectives [24].

Recycling Rate Per Sector	2015	2025	2030
Plastics	25%	50%	70%
Cardboard and Paper	27%	50%	80%
Metals	46%	60%	80%
Used oil	36%	50%	70%
WEEE	12%	20%	40%
Batteries	30%	50%	80%
Tires	42%	60%	80%

4. Comprehensive Analysis of Municipal Solid Waste-Management Systems

4.1. Legal and Financial Frameworks in Target MENA Region Countries

One of the critical components that maintains the operation of highly advanced solid waste management systems is the government's ability to implement existing policies and regulations as well as develop and issue new regulations and laws in line with the requirements of the transition towards adopting sustainable solutions within the framework of waste management and the circular economy. Below is a detailed description of the most prominent laws and regulations covering waste management in the targeted countries in the MENA region.

4.1.1. Algeria

Since the 2000s, an important legal directive has been put in place to allow Algeria to respect the international commitments that the country has made and to ensure that environmental issues are addressed from a sustainable development perspective.

In Algeria, the task of municipal solid waste collection has been delegated to the municipalities by Law No. 01-19 of 12 December 2001 relating to waste management, control, and disposal. This law defines the basic principles that lead to the integrated management of waste, starting with its generation, followed by its collection, and ending with its treatment and disposal. Table 4 summarizes the Algerian legislative acts for municipal solid waste management [27].

In the waste management sector, Law 01-19 has been reinforced by several executive texts taking into account the following principles:

- Prevention of and reduction in the production and harmfulness of waste at the source;
- The organization of sorting, collection, transport, and treatment of waste;
- The recovery of waste by reuse, recycling, and any other action aimed at obtaining, from this waste, reusable materials or energy;
- The environmentally sound treatment of waste;
- Informing and raising citizens' awareness of the risks presented by waste and their impact on health and the environment, as well as the measures taken to prevent, reduce, or compensate for these risks [38].

Table 4. Algerian laws for municipal solid waste management [27].

Law and Executive Decrees	The Related Field
Law N°. 01-19	The management, control, and disposal of waste.
Law N°. 03-10	The protection of the environment in the context of sustainable development.
Executive decree N°. 02-175	The creation of the national waste agency.
Executive decree N°. 04-410	The general rules for the development and operation of waste treatment facilities and the admission of waste at these facilities.
Executive decree N°. 07-205	The modalities and procedures for the preparation, publication, and revision of the scheme of municipal household and similar waste management.
Executive decree N°. 04-199	The modalities for the establishment, organization, operation, and financing of the public system of treatment and recovery of packaging waste.

Several actors at both national and local levels are directly involved in waste management in Algeria. Among these are:

- The Ministry of the Environment through its various instruments, in particular the National Waste Agency, whose main mission is the accompaniment and support of organizations active in the waste management sector, in particular local communities (municipalities, collection operators, management institutions, etc.).
- The Ministry of the Interior, Local Authorities, and Regional Planning, which provides financial and logistical support to the Communal People's Assemblies by making annual grants. The amount reserved for urban waste management is quite significant.

Other ministries are also involved in the field of waste management. These include the Ministry of Health (dealing with medical waste), the Ministry of Industry (dealing with special hazardous waste), and the Ministry of Agriculture (dealing with agriculture waste and phytosanitary). Primarily, municipal waste management is the responsibility of local authority cleaning services (APC). This makes it difficult to control and monitor all phases of municipal solid waste: collection, transportation, treatment, and disposal of waste. In each city, a private company affiliated with the local authority is established, which is responsible for waste management in the concerned city, including the collection of fees. The primary role of local authorities is to supervise and monitor the performance of the private companies and ensure the cleanliness of cities. Despite the efforts made by the government, the participation of local communities in the field of waste management is still very limited. However, the average fee per person for household waste management (mainly collection and disposal) ranges from 4 to 7 USD/year [14].

4.1.2. Egypt

Egypt generates around 26 million tons of MSW annually, with generation rate varying significantly across the country. It has ambitious plans for improving the current MSW management systems [15]. To achieve these plans, numerous infrastructure waste management projects are being initiated, and a regulatory framework has been established. Law 202 for 2020 promulgating the Waste Management Regulation Law was issued on 13 October 2020 by the Egyptian Ministry of Environment. However, the executive regulation of the law has not yet been released. The law aims to develop the integrated management of municipal, industrial, agricultural, demolition, and construction waste as well as their safe disposal. The law also aims to reduce waste generation, promote reuse, and ensure the recycling, treatment, and final disposal of waste, and finally, to manage waste in a way that reduces damage to public health and the environment [18]. Concerning MSW, the new law does not include any articles calling for mandatory source separation nor prohibiting the landfilling of any specific waste streams or materials.

The financial instruments applied for MSW management in Egypt involve a type of pay-as-you-throw fee. However, the fee is based on an estimated waste-generation rate that varies from one neighborhood to another across the different Egyptian cities and governorates where the MSW generation varies from 0.3 to 2.0 kg/capita/day. The new law states, in Article 34, that this fee ranges between 2 and 40 EGP per month per residential unit, 30 to 100 EGP for commercial units, up to 5000 EGP for governmental and community services establishments, and up to 20,000 EGP for commercial, industrial, and touristic establishments. Residential and commercial units are identified according to the number of electricity meters and, accordingly, the fees are collected with the electricity bills. Nevertheless, the fees collected only partly cover the waste management services' financial needs, which demand considerable municipal spending of approximately 7.2 billion EGP per year. Consequently, the new law, in Article 16, calls for extending the producers' responsibility to cover the end-of-life environmentally sound management and disposal costs of their products; yet, according to Article 17, the government shall issue a decree identifying priority products to be managed according to an Extended Producer Responsibility (EPR) financial scheme [18]. This has been proposed as a solution for the current financial challenges of waste management. The proposal relates to three waste streams that are extensively present in MSW, which are tires, packaging, and waste electrical and electronic equipment (WEEE) (Sustainable Recycling Industries, 2017). Furthermore, currently, the Egyptian government are working on establishing a strategy for EPR implementation for packaging waste, considering the different packaging materials available.

4.1.3. Jordan

Solid waste management within the urban and rural areas of the country is mainly implemented by municipalities. On the other hand, the disposal process into the sanitary landfills and dump sites (excluding Greater Amman Municipality) is undertaken by Joint Services Councils (JSCs), who own and operate 16 landfills. The JSCs are governmental agencies that are formed from several member municipalities located within one geographical region. They mainly deal with solid waste disposal [34,40].

Until recently, Jordan had no special bylaw that regulated the solid waste generated in urban and rural centers of the country. However, in 2020, the Jordanian parliament passed the Waste Management Framework Law for the year 2020. This identified five basic principles of solid waste management, namely, prevention, the precautionary principle, extended producer responsibility, the polluter pays principle, and the proximity principle. Before passing the law, the Jordanian government endorsed the National Solid Waste Management Strategy, which established a road map to shift from a disposal driven (end of pipeline) solid waste management system into a more sustainable integrated system based on the reduce, reuse, and recycle approach.

The main institutions that are responsible for solid waste management include the Ministry of Environment (MoEnv), which is mainly concerned with the regulatory and policy aspects as well as monitoring of enforcement and compliance with these regulations. The Ministry of Local Administration (MOLA) is responsible for the supervision of the municipal functions and providing them with financial support. On the other hand, municipalities are responsible for the day-to-day work of solid waste management within their boundaries, which includes collection, transport, and transfer. Street sweeping is also part of the municipalities' work. The cemeteries are managed and operated by the JSCs. These are a consortium of a group of municipalities located in one geographic area and share the use of landfills to dispose of the solid waste generated by them [34].

4.1.4. Morocco

The primary objective of sustainable solid waste management is to address concerns related to public health, environmental pollution, land use, resource management, and the social and economic impacts associated with improper waste disposal. The main challenges facing the municipal solid waste management system in Morocco are the lack of appropriate infrastructure and adequate funding, especially in remote and small areas. Against these challenges, Morocco has taken several steps, engaged in an ambitious sustainable development agenda, and undertaken a wide range of reforms. Among those are the National Charter for Environment and Sustainable Development, the National Strategy for Sustainable Development 2015–2030, and the Nationally Determined Participation Commitments (INDC) through which Morocco commits to a 32% reduction in greenhouse gas emissions (GHG) by 2030. The framework is Law No. 99-12 on the National Charter for Environment and Sustainable Development. Morocco, like many other countries, has embarked on multiple regulatory and institutional approaches in this field of waste management by strengthening the legal arsenal that will serve as a framework for public authorities in this field. Law No. 28-00 on Waste Management and Disposal was promulgated to regulate waste management by covering the entire chain from collection to disposal, including treatment and recovery. For example, the Ministry of Energy, Mines and Environment is responsible for, among other treatments, treating waste by anaerobic fermentation method. The purpose of the law is to lay the foundations for a waste management policy, which in turn ensures the improvement of the performance of the management processes in force in addition to minimizing, as much as possible, the negative effects of waste on human health and the environment. The Moroccan Waste Code distinguishes between municipal waste, non-hazardous industrial waste, medical waste, pharmaceutical waste, agricultural waste, and hazardous waste, each of which falls under different planning and treatment regimes. Indeed, there are many laws and regulations in Morocco that regulate various aspects of waste management. Law 11-03 deals with the protection and development of the environment, while Law 12-03 focuses primarily on environmental-impact studies. A regulatory text for waste incineration activity was approved through Decree No. 2-12-172 issued in 2012, which specifies technical instructions related to waste removal and recovery operations by incineration. The disposal of plastic bags and their replacement with biodegradable bags is regulated in accordance with Law No. 22-10. Morocco consumes 26 billion plastic bags every year and the quota per capita is about 900 bags per year. In July 2016, Law No. 75-15 called "Zero Mika" came into force "which prohibits the manufacture, import, export, marketing, and use of plastic bags". On the other hand, Decree No. 2-09-139 regulates the management of medical and pharmaceutical waste in terms of the rules for sorting, packaging, collection, storage, transportation, and disposal. In Morocco, it is apparent that the actions implemented relate, in large part, to the use of landfill of waste in the absence of any approach aimed at reducing their production. Current practices for the implementation of the national municipal waste policy are represented mainly by the National Household Waste Program (PNDM). The Law 28-00 created an obligation for public inquiries for master, provincial, and landfill plans (Decree n° 2-09-683, 2010). It sets out the modalities for the development of the regional master plan for the management of industrial, medical, pharmaceutical, non-hazardous, and agricultural waste. The National Waste Management Program is part of the policy of reforming and developing the municipal waste sector, which aims to structure this sector in local plans, especially by improving waste collection services, establishing waste treatment and recovery centers, and closing and rehabilitating old landfills. For hazardous waste, and in accordance with Article 9 of Law 28-00, a National Master Plan for Hazardous Waste (PDNDD) is being finalized in order to create an integrated and sustainable management system for hazardous waste. The National Charter for Environment and Sustainable Development (CNEDD) launched in 2009 consolidated the achievements of civil society participation at the local level and made it an important part of implementing the rule of law and legislation. After reviewing the most prominent laws regulating waste management in the target countries and in order to assess the situation more accurately, Table 5 summarizes the extent to which targeted MENA region countries apply the most prominent laws in force adopted by developed countries in the field of waste management and the circular economy.

Table 5. The extent to which the targeted MENA region countries apply the most prominent laws in force adopted by developed countries.

Regulatory Instruments	Targeted MENA Region Countries			
	Algeria	Egypt	Jordan	Morocco
Bans and Restrictions (e.g., a landfill ban)	N/A	N/A	N/A	N/A
Mandatory Source Separation	N/A	N/A	N/A	N/A
Economic Instruments				
Extended Producer Responsibility (EPR)	N/A	Planning phase	Working on it	N/A
Deposit-Refund	N/A	N/A	N/A	N/A
Landfill/Incineration Tax	N/A	N/A	Minimal	N/A
Pay-As-You-Throw (PAYT)	N/A	Partially applied	In Amman only. Flat rate in other cities	N/A

N/A: Not Available

It is clear from Table 5 that the region suffers from a lack of many laws and regulations, which, if they existed, would have contributed significantly to a qualitative leap in the waste management systems currently in use. Indeed, the first step that precedes the implementation of any waste management strategy or plan is the enactment of laws that support that strategy and ensure its success. Developed countries would not have reached their current advanced levels in the field of waste management and the circular economy without effective and strict laws. For example, Germany passed a law in June 2005 stating that waste can no longer be landfilled without pre-treatment, which in turn contributed to the transition from waste disposal to waste management. Reaping the fruits of this transformation took nearly 20 years of development and research. Therefore, all current laws in the region must be laid out for review and discussion with decision-makers in order to ascertain what is appropriate and to develop new laws that are compatible with the MENA region's circumstances and requirements.

4.2. Waste Sorting and Collection

Unlike high-income developed countries where there is public and societal awareness of waste sorting, sorting activities at the household level in developing countries, some of which are MENA region countries, are still limited and almost non-existent in some countries. Therefore, MSW generally consists of mixed waste containing food and other types of waste. Waste sorting is usually performed by poor families or the so-called informal sector in order to earn additional income from selling recyclable materials. Despite the high proportion of the municipal budget being spent on waste collection, about 80–90% of the total municipal solid waste budget, the efficiency of municipal solid waste collection is still very low in many countries, and this varies within the country itself. For example, in Jordan, the collection rate in Amman reaches 90%, while in remote or rural areas, the efficiency of the collection process stands at 50%, and sometimes less. Moreover, due to the low efficiency of waste collection, dumping of waste on the side of the road is a common practice and sometimes some of the public take up the task of burning waste in public places and on the roadside [40].

4.3. Waste Treatment and Disposal

Open dumping and landfilling are the most common methods of MSW disposal in developing countries, mainly because they are the cheapest treatment methods especially when social and environmental impacts are not considered. Together, these two methods account for about 70–90% of the total municipal solid waste. Compared with other treatment methods, open dumping and landfilling pose the highest risks to the environment and human health, causing deterioration in soil and water quality, air pollution, and the spread of diseases [34]. Recycling and material recovery activities are still in their initial stages due to the absence of supporting policies, regulations, and laws, in addition to the absence of public awareness of the importance of separating materials from the source. Incineration and energy recovery are very limited due to the high cost of investment and inappropriate composition of mostly inert and biodegradable waste that has high water content, which negatively affects the calorific value of waste. Composting has long been promoted as a method of treating biodegradable waste, creating new job opportunities and additional income generation for communities. However, many problems arise for composting practices. Composting run by municipalities often faces technical problems due to a lack of experience and know-how as well as the use of mixed MSW, which produces poor quality compost. Therefore, the application of composting practices is still limited to small or experimental projects, which are often unable to produce a market-oriented end product.

Despite most MENA region countries starting to improve their waste management services and adopting new strategies to achieve a sustainable waste management system by applying some advanced environmental technologies and expanding the scope of recycling and reuse, success is still limited, as the percentage of waste that is landfilled is still at its highest levels. This is clearly obvious from Table 6, which summarizes the practices adopted by the four countries covered by this study.

Table 6. Summary of the MSW practices adopted by the MENA region countries.

Country	Million Tons y^{-1}	%	Treatment Method				
			Landfilling Rate (%)	Recycling Rate (%)	Composting Rate (%)	Anaerobic Rate (%)	WtE Rate (%)
Algeria	13.5	100	92	7	1	-	N/A
Egypt	26	100	88	12	Incl. in recycling	N/A	N/A
Jordan	3.2	100	90	10	-	-	-
Morocco	7.4	100	91	9	NA	NA	NA

The typical problems in municipal solid waste management in the region can be identified as insufficient collection systems, limited use of recycling and energy-recovery activities, and lack of know-how and skilled manpower.

Moreover, an overview of solid waste management practices in the four selected MENA countries is provided in Table 7. It is quite clear that there are many problems in the handling, collection, transportation, treatment, and disposal of solid waste in the MENA region countries examined.

Table 7. Overview of MSW practices in the MENA region countries.

Activity	MENA Region Countries			
	Algeria	Egypt	Jordan	Morocco
Source reduction	Discussions about source reduction, but it is rarely incorporated into any organized plan.	Unregulated, some discussions about source reduction, but it is rarely incorporated into any organized plan.	Discussions about source reduction, but it is rarely incorporated into any organized plan.	Structured educational programs began to emphasize source reduction and material reuse.
Collection	Improving service and increasing collection from residential areas. Larger vehicle fleet and more mechanization.	Intermittent and inefficient. Improving service and increasing collection from residential areas. Collection rate up to >90%. Larger vehicle fleet. Compactor trucks and highly mechanized vehicles are common.	Collection rate up to >90%. Compactor trucks and highly mechanized vehicles are common.	Improving service and increasing collection from residential areas. Collection rate up to >90%. Larger vehicle fleet. Compactor trucks and highly mechanized vehicles are common.

Table 7. Cont.

Activity	MENA Region Countries			
	Algeria	Egypt	Jordan	Morocco
Recycling	The informal sector is still involved. Some high-tech sorting and treatment facilities. The material is often imported for recycling.	Most of the recycling is done by the informal sector and waste pickers. Some high-tech sorting and treatment facilities. The material is often imported for recycling. Increased interest in long-term markets.	Most of the recycling is done by the informal sector and waste pickers.	The informal sector is still involved. Some high-tech sorting and treatment facilities. The material is often imported for recycling.
Composting	It is rarely conducted formally even though the waste stream has a high organic matter content.	It is rarely conducted formally even though the waste stream has a high organic matter content. Large composting plants are usually unsuccessful; small facilities are more sustainable.	It is rarely conducted formally even though the waste stream has a high organic matter content.	Large composting plants are usually unsuccessful; some small composting facilities are more sustainable.
Incineration	Some incinerators are used, but they face financial and operational problems. It is not as common as in high-income countries.	They are not popular or successful due to the high CAPEX and OPEX costs, high moisture content of the waste, and high proportion of inert materials. Some incinerators are used, but they face financial and operational problems. It is not as common as in high-income countries. -Predominant in areas where land costs are high. Most incinerators equipped with environmental controls and energy-recovery systems.	For medical waste only.	Some incinerators are used, but they face financial and operational problems. It is not as common as in high-income countries.
Landfilling	Sanitary landfills equipped with a combination of linings, leak detection, leachate collection, and treatment systems.	Low-tech sites. Open dumping is still the most practiced method. Some controlled and sanitary landfills equipped with a combination of linings, leak detection, leachate collection and treatment systems.	About 70% of the solid waste generated goes to sanitary landfills, while the remainder is disposed into unsanitary landfills.	Some controlled and sanitary landfills with some environmental controls. Open dumping is still the most practiced method.
Costs	Collection costs account for 80–90% of the MSWM budget. Fees for waste management are regulated by local governments. The fee collection system is inefficient.	Collection costs account for 80–90% of the MSWM budget. Fees for waste management are regulated by local governments. The fee collection system is inefficient.	Collection costs account for 80–90% of the MSWM budget. Fees for waste management are regulated by local governments. The fee collection system is inefficient.	Collection costs account for 80–90% of the MSWM budget. Fees for waste management are regulated by local governments. The fee collection system is inefficient.

The waste management system in most MENA countries is limited to collection and transportation. During the past few decades, there has been a remarkable development in the solid waste collection and transportation system, but the recovery of materials and energy is still in its initial stages. Although these countries have tried to move towards a sustainable waste management system, their attempts have not brought about any lasting change as they were basically pilot initiatives or projects that ended when donor funding expired. This is due to the absence of planning, the use of inappropriate technology, and the inability of the municipalities to continue in light of their financial deficits. This indicates the urgent need to shift from poor management of municipal solid waste to a more effective and sustainable management system.

5. Outlook from Selected MENA Region Countries

5.1. Algeria

The general approach to solid waste management should be based on strengthening legislative and institutional frameworks, as is the case in developed countries. This includes public–private partnership (PPP) in waste management, and environmental, technical, and economic aspects of the processes used for treatment and recovery, in parallel with the adoption of a permanent education and awareness policy on waste management [26,41]. The management of urban solid waste is part of the National Action Plan for Environment and Sustainable Development (PNAE-DD) through the adoption of a National Program for Integrated Management of household waste (PROGDEM). It aims for an integrated, phased, and progressive management approach to household waste [41]. This program has defined the main directions for the implementation of this management through:

- The reorganization of the municipal administration responsible for waste management;
- Develop legislation and laws regulating waste management that promote recycling, energy and materials recovery, and reduce amount waste that ends up in landfills;
- Capacity building in the field of waste collection and transportation services;
- Providing an opportunity for private investment in the field of waste management and related public services;
- Implementation of training and technical assistance programs;
- Establish cooperation between the private and public sectors, which is required in order to ensure the technical, financial and social sustainability of the implemented solid waste management system.

5.2. Egypt

Egypt has ambitious plans to improve MSW management systems that are listed in Egypt Vision 2030. These involve increasing the waste collection coverage and efficiency from 20% and 60%, respectively, in 2016, to reach 80% and 90%, respectively, by 2030. The plans also include increasing the recycling rate to 25% by the year 2030, as well as ensuring 100% safe disposal of hazardous wastes [28,42] (EEAA, 2018; MPED, 2016). Furthermore, WMRA has been working since 2018 towards increasing the collected MSW recycling rates to 80% by 2026. Of this, 60% will be recycled for compost and RDF production, and 20% will be thermally treated for energy production. These plans are reinforced by numerous waste management infrastructure projects to build transfer stations, recycling plants, and sanitary landfills, as well as cleaning of waste accumulation sites, to cover the country completely in a four-year plan starting in 2019 [15,33]. The RDF production is being promoted by the Egyptian government, as the Ministerial Decree No. (49) of 2021 obliges cement plants to use a minimum of 10% RDF in the alternative fuels used for their energy requirements. The decree comes in response to the Ministry of Environment's plans to increase the waste-to-energy practices in cement plants to 15% by 2030 through the utilization of nearly 22 million tons of solid waste and 30 million tons of agricultural residues to produce RDF [21]. Moreover, waste-to-energy projects are being planned, relying on MSW through incineration, pyrolysis, and gasification. Minimization of single-use plastics is also currently considered to be an MSW management focus area, as a response to the extremely high consumption of plastic bags in Egypt, which accounts for around 12 billion bags annually littering Egypt's streets and waterways [43]. Furthermore, numerous consultancy projects are currently being conducted to set strategies, guidelines, and roadmaps for improving MSW management through dumpsite closure, packaging-waste extended producer responsibility, and improving the management of other waste streams that are often present in the Egyptian MSW such as medical, electronic, construction, and demolition waste. These MSW management developments will be complemented with some important pillars including:

Introducing better definitions for waste management, practices, recycling, and recovery, to ensure that all the stakeholders are in agreement.

Inclusion of all the relevant formal and informal stakeholders in the formal development and plans for MSW management and ensuring their proper understanding of all the relevant regulations as well as national and international codes of design and practice.

Promoting MSW source separation, as well as MSW waste stream separation from other waste streams commonly mixed with MSW through its management chain such as medical, industrial, construction, and demolition waste.

Developing business models for waste management interventions to encourage private-sector investment in MSW management.

5.3. Jordan

The national solid waste management strategy that was adopted by the Jordanian government in 2015 has established a road map for moving from the traditional end-of-pipeline technology towards a more sustainable and integrated solid waste management system [44].

Table 8 explains the proposed composting plants and their location and capacities according to the national solid waste management strategy.

Table 8. Proposed composting plants and their location and capacities according to the national solid waste management strategy [34,44].

Region	Governorate/ Municipality	Design Capacity (ty^{-1})	Initial Starting Amount (ty^{-1})	Year of Operation
Northern	Irbid	84,000	21,000	2025
	Mafraq	22,000	5500	2025
Central	Amman & Zarqa	215,000	53,750	2025
	Salt Greater municipality	25,000	6250	2025
Southern	Aqaba	25,000	6250	2025

During the short and mid-term periods (until the year 2024) the strategy aims to convert the existing landfills into sanitary landfill sites. On the other hand, by the end of the long-term period (2025–2034), there should be the following facilities in operation:

- Three major clean material recovery facilities (MRFs);
- Two mechanical biological treatment facilities for mixed waste;
- Two anaerobic digestion units;
- Five composting plants.

5.4. Morocco

The national strategy predicts that the total amount of waste that will be landfilled by 2030 is about 39 million tons. The implementation of this strategy should reduce waste generation, increase material and energy recovery, and reduce environmental degradation associated with waste management. The cost was estimated at USD 0.5 billion or 0.4% of GDP. In order to encourage all companies involved in the waste management sector and the circular economy, the General Confederation of Moroccan Enterprises (CGEM) created in 2016 the Coalition for the Valorization of Waste (COVAD). An entrepreneurial dynamic was publicized on the occasion of the COP22, in particular via the Moroccan platform "Initiatives Climat", which identified and recounted personalized success stories. Two sectors were highlighted: anaerobic digestion and the production of organic biofuels.

In addition, Morocco introduced its first environmental tax (the ecotax) in 2014, consisting of 1.5% based on value when selling and importing plastics. Imposing this was not without struggle with the plastics industry. However, its proceeds now bring in USD 17 million to the National Environment Fund, which aims to finance the development of new recycling channels. Morocco highlighted its dedication to environmental protection and sustainability in the 2021 agreement of the United Nations Environment Assembly (UNEA), underlining its commitment to global cooperation in the name of a "green world". Morocco remains highly focused on environmental sustainability and the transition to green energy. The policy boasts a comprehensive plan to push Morocco into a "green economy" by 2030 through targeted investments, subsidies, and reforms. In Washington DC on 15 December

2020, the World Bank Board of Executive Directors approved a USD 250 million (EUR 214.2 million) program to support the Moroccan agriculture sector's Green Generation Strategy, as part of a joint operation with the French Development Agency (AFD). The Green Generation Program-for-Results is designed to make farming more rewarding and strengthen sustainable agriculture by streamlining climate-smart practices.

6. Summary

Waste management in the MENA region is an extremely critical issue. Some years ago, significant changes occurred in the operational environment of the waste sector in the region due to the Arab Spring, which led to massive and sudden migration in many countries and a large influx of refugees. Jordan is one of those countries, for example. This has led to the creation of an additional environmental and economic problem represented by increasing pressure on infrastructure, greatly increased pollution, and rapid urbanization, which negatively affects the waste management system in the host countries.

However, this work has shown that some countries have already begun to improve their waste management service and have adopted new strategies to achieve a sustainable system within the framework of waste management and the circular economy. Indeed, some countries have implemented some advanced environmental technologies and expanded the scope of recycling and reuse. Nevertheless, this progress has been quite limited and unsustainable, and did not go beyond the level of pilot projects, which often ended when funding expired.

On the other hand, this work shows that the solid waste sectors in the MENA region suffer from critical problems and face many challenges in terms of waste management and the techniques applied to improve it. The typical problems in MSW management in the region can be identified as the increase in per capita MSW generation rates, the gaps in current related legislation, financial constraints, lack of know-how, limited material and energy-recovery activities, insufficient collection systems, and lack of trained and skilled manpower. Therefore, it was clear from the study that waste management challenges in the region are common and similar. They can be summarized, but are not limited, as follows:

The lack of legislation and weak implementation are considered two of the main challenges facing waste management in the MENA region countries. Furthermore, the lack of tools to assess and monitor the performance of waste treatment technologies and their output quality and utilization is considered one of the main problems.

The low level of cooperation between national institutions and local authorities, and the overlapping roles. Governments should define roles, prioritize responsibilities, and develop regulations to ensure a harmonized framework.

There is an urgent need to review and reform the waste management sector to improve its services in many countries in the region. Most waste sectors in the region have central waste management in terms of planning and defining tariffs, regulations, and laws that regulate waste management systems, which can be developed by decentralizing these sectors and giving municipalities powers that enable them to provide a better service.

The impact of refugee flows due to political instability in the Middle East and North Africa region, which has burdened the waste sectors and doubled the pressure on the infrastructure.

The limited participation of the national private sector in decision-making and developing plans and strategies in waste management in most countries of the Middle East and North Africa. There is an urgent need in most MENA countries for private-sector participation and investment in the waste management sector to achieve the sustainable development goals.

Limited funding sources and their negative impact on the quality of waste management and the sustainability of projects in the region. The fees collected for managing waste are very low, covering no more than 30% of the costs. The funding system for waste management is unequally distributed, as 90% of the funds available go to logistics and only 10% remains for treatment. Therefore, work, should be done to strengthen and empower the municipalities so that they are able to develop and implement sustainable solutions that are compatible with their circumstances and capabilities.

Separate collection is still not practiced in the region. The MSW is collected as mixed waste and contains hazardous materials. Up to 90% of it is sent to different controlled landfills and dumping sites without any treatment, which generates high levels of methane gas due to the high amount of organic material and water content. The rate of recyclable material recovered from the waste stream is about 5–10%.

Indeed, most of the approaches undertaken in the past have included the adoption of treatment and disposal technologies without a full understanding of the process, which has led to the lack of knowledge and local skilled manpower. All of these issues emphasize the weaknesses and problems in the previous measures. In order to overcome these problems, first, a technical solution should be developed with the required modifications and proper implementation considering the legal, financial, and management aspects.

7. Recommendations

The governments must promote an integrated solid waste management hierarchy and set up a national policy regarding the minimization of waste to landfill by making a shift from waste disposal towards waste management and the circular economy.

Attention should be paid to increasing coordination and linking governmental bodies, institutions, and services at the administrative and legislative levels to improve management in terms of waste collection, treatment, and disposal.

Establish more cooperation between local waste management authorities and different ministries responsible for various policies including waste management, energy, and the environment in order to explore the full potential of implementing a sustainable waste management system that adopts a holistic approach.

It is essential that local industry and educational institutions have regular links with waste management regulatory bodies. This can be facilitated by creating a network platform that brings together all stakeholders in the waste management sector.

A network concerned with waste management should be created at the level of the countries of the MENA region for the various member states to share their experiences in the research and development of innovative technology to reveal the potential in the recycling industry and convert waste into energy and any regulatory framework to accommodate it.

Reconsider the waste management fee system through the establishment of the extended product responsibility system.

Establish cooperation between the private and public sectors involved in the solid waste management system and the circular economy in order to ensure the technical, financial, and social sustainability of the solid waste management system implemented.

Cooperate with universities and research institutions in order to support municipalities facing challenges, provide sustainable solutions, and develop plans to improve the waste management system and provide better services.

Launch large-scale awareness programs on solid waste management and the circular economy for all citizens and stakeholders. Work on developing preventive measures represented by adopting environmentally responsible behaviors and attitudes that make it possible to carry out sorting, selective collection, and recovery of materials and energy from waste.

8. Conclusions

Solid waste management has become one of the major environmental problems facing municipal authorities in the MENA region. It has been aggravated over the past few years by the sharp increase in the volume of waste generated as well as qualitative changes in its composition. The provision of adequate waste management services is critical because of the potential impact on public health and on the environment. Lack of planning, lack of proper disposal, insufficient collection services, use of inappropriate technology, inadequate financing, and limited availability of trained and qualified manpower together with massive and sudden population increases are considered to be the main problems facing solid waste management in the region.

The aim of this study was to identify the different practices and approaches of solid waste management employed in selected MENA region countries, and the extent to which the policies, regulations, and technologies applied play any role in the context of solid waste management and the circular economy.

The study revealed that most waste management issues in the countries analyzed appear to be due to political factors and the decentralized nature of waste management with multi-level management and responsibilities. Material and energy recovery in the context of municipal solid waste management does not differ significantly in the countries in the MENA region considered; in most cases, "waste" is still seen as "trouble" rather than a resource. Therefore, a fresh look is required and there is a need for a paradigm shift from a linear economy of "waste management", to a circular-economy model of "resource efficiency". The latter can be achieved by adopting a joint vision on how the solid waste management system can be transformed into a circular economy. Since they are the main drivers of any transition towards a circular-economy system, the policies, strategies, and practices in place that regulate the performance of waste management in the region must be revised.

For fairness, despite the financial and technical obstacles facing the waste management sector, continuous attempts are made to divert waste from landfills to some advanced recycling facilities. Indeed, most countries in the MENA region are approaching sustainable waste management solutions and most decision makers do identify their problems and previous mistakes for setting a solid waste management system. Currently, there is a consensus among the concerned authorities in the region that the success of any sustainable waste management solution requires the involvement and cooperation of all parties involved in the sector such as international companies, local private companies, and municipalities. Awareness is also being raised regarding the necessity of shifting towards the implementation of waste management in the context of a circular economy.

Author Contributions: Conceptualization, S.H. and O.A.; methodology, S.H.; validation, S.H., O.A., L.E.F., S.E., I.H. and J.G.; investigation, S.H., O.A., L.E.F. and S.E.; resources, S.H., O.A., L.E.F., S.E., M.H., B.C. and M.A.; data curation, S.H. and O.A.; writing—original draft preparation, S.H., O.A., L.E.F. and S.E; writing—review and editing S.H.; visualization, S.H.; supervision, S.H.; project administration, S.H., I.H. and J.G.; funding acquisition, S.H. All authors have read and agreed to the published version of the manuscript.

Funding: This research funded by the PREVENT Waste Alliance, an initiative of the German Federal Government in the framework of the German–MENA University Network for Waste Management and Circular Economy project (Grant number 81263215).

Data Availability Statement: Not applicable.

Acknowledgments: This research was carried out within the framework of the German–MENA university network for waste management and circular economy project, which is funded by the PREVENT Waste Alliance, an initiative of the German Federal Government.

Conflicts of Interest: The authors declare no conflict of interest.

References

1. EIA. *Countries Profile, Middle East and North Africa*; U.S. Energy Information Administration: Washington, DC, USA, 2015.
2. World Bank. Population, Total-Middle East and North Africa. 2020. Available online: https://data.worldbank.org/indicator/SP.POP.TOTL?locations=ZQ (accessed on 20 October 2021).
3. Onwosi, C.; Igbokwe, V.; Odimba, J.; Eke, I.; Nwankwoala, M.; Iroh, I.; Ezeogu, L. Composting technology in waste stabilization: On the methods, challenges and future prospects. *J. Environ. Manag.* **2017**, *190*, 140–157. [CrossRef] [PubMed]
4. Negm, A.M.; Shareef, N. (Eds.) Introduction to the Waste Management in MENA Regions. In *Waste Management in MENA Regions*; Springer Water; Springer International Publishing: Cham, Switzerland, 2020; pp. 1–11. ISBN 978-3-030-18350-9.
5. Hemidat, S. Feasibility Assessment of Waste Management and Treatment in Jordan. Ph.D. Thesis, The Academic Board of Rostock University, Rostock, Germany, 2019.

6. Abu-Qdais, H.; Gibellini, S.; Vaccari, M. Managing Solid Waste under Crisis: The Case of Syrian Refugees in Northern Jordan. In Proceedings of the Sardinia 2017—Sixteenth International Waste Management and Landfill Symposium, Cagliari, Italy, 2–6 October 2017.
7. Nassour, A.; Al-Ahmad, M.; Elnaas, A.; Nelles, M. Practice of waste management in the Arab region. Contribution in: Water-to-resources 2011-4. In Proceedings of the International Conference MBT and Sorting Systems, Hanover, Germany, 24–26 May 2011; Kule-weidemeier, M., Ed.; pp. 81–91.
8. Abu-Qdais, H.; Wuensh, C.; Dornack, C.; Nassour, A. The role of solid waste composting in mitigating climate change in Jordan. *Waste Manag. Res.* **2019**, *37*, 833–842. [CrossRef] [PubMed]
9. Hemidat, S.; Saidan, M.; Al-Zu'bi, S.; Irshidat, M.; Nassour, A.; Nelles, M. Potential Utilization of RDF as an Alternative Fuel for the Cement Industry in Jordan. *Sustainability* **2019**, *11*, 5819. [CrossRef]
10. Karak, T.; Bhagat, R.M.; Bhattacharyya, P. Municipal solid waste generation, composition, and management: The world scenario. *Crit. Rev. Environ. Sci. Technol.* **2012**, *42*, 1509–1630. [CrossRef]
11. Chaher, N.E.H.; Chakchouk, M.; Nassour, A.; Nelles, M.; Hamdi, M. Potential of windrow food and green waste composting in Tunisia. *Environ. Sci. Pollut. Res.* **2020**, *28*, 46540–46552. [CrossRef] [PubMed]
12. Chaher, N.E.H.; Hemidat, S.; Thabit, Q.; Chakchouk, M.; Nassour, A.; Hamdi, M.; Nelles, M. Potential of Sustainable Concept for Handling Organic Waste in Tunisia. *Sustainability* **2020**, *12*, 8167. [CrossRef]
13. Boudghene Stambouli, A. Algerian renewable energy assessment: The challenge of sustainability. *Energy Policy* **2011**, *39*, 4507–4519. [CrossRef]
14. AND—Agence National des Déchets (National Waste Agency). *2020 Rapport sur L'état de la Gestion des Déchets en Algérie*; AND: Algiers, Algeria, 2020.
15. MoE—Ministry of Environment. *Municipal Solid Waste Management System "Infrastructure"*; MoE: Cairo, Egypt, 2021.
16. EEAA. *Egypt State of the Environment Report*; EEAA: Chatswood, Australia, 2017.
17. CAPMAS. Population of Egypt. Central Agency for Public Mobilization and Statistics. 13 July 2021. Available online: https://www.capmas.gov.eg/Pages/populationClock.aspx# (accessed on 25 September 2021).
18. MoE—Ministry of Environment. *Law 202 for 2020 Promulgating the Waste Management Regulation Law*; MoE: Cairo, Egypt, 2020.
19. Abdallah, M.; Arab, M.; Shabib, A.; El-Sherbiny, R.; El-Sheltawy, S. Characterization and sustainable management strategies of municipal solid waste in Egypt. *Clean Technol. Environ. Policy* **2020**, *22*, 1371–1383. [CrossRef]
20. Enterprise. *Where did Things go Wrong with Egypt's Waste Management?* Enterprise: Aliso Viejo, CA, USA, 2020.
21. CemNet. Egypt: Cement Plants to Use 15% of Waste by 2030. 2016. Available online: https://www.cemnet.com/News/story/160444/egypt-cement-plants-to-use-15-of-waste-by-2030.html (accessed on 12 September 2021).
22. Gibellini, S.; Abu Qdais, H.; Vaccari, M. Municipal solid waste management in refugee hosting communities: Analysis of a case study in northern Jordan. *Waste Manag. Res.* **2021**, 0734242X21994656. [CrossRef] [PubMed]
23. DOS—Department of Statistics. *Jordan in Figures*; Department of Statistics: Amman, Jordan, 2017; pp. 33–34.
24. SNRVD. National Waste Reduction and Recovery Strategy. Summary Report. 2019. Available online: https://www.logipro.ma/images/Traitement_des_deee/Rapport_de_synthese_SNRVD_FR.pdf (accessed on 10 August 2021).
25. UNEP. *Integrated Waste Management Scoreboard—A tool to Measure Performance in Municipal Solid Waste Management*; UNEP: Nairobi, Kenya, 2005.
26. Djemaci, B.; Ahmed Zaïd, M. *Integrated Solid Waste Management in Algeria. Constraints and Limits of Its Implementation*; CIRIEC Act 12; CIRIEC: Liège, Belgium, 2011.
27. Kouloughli, S.; Kanfoud, S. Municipal Solid Waste Management in Constantine, Algeria. *J. Geosci. Environ. Prot.* **2017**, *5*, 85–93. [CrossRef]
28. EEAA. *State of the Environment 2017–Summary for Policymakers*; EEAA: Chatswood, Australia, 2018.
29. UNEP. *National Action Plan for Sustainable Consumption and Production (SCP) in Egypt*; UNEP: Nairobi, Kenya, 2016.
30. African Union. *AGENDA 2063: The Africa We Want-First Ten-Year Implementation Plan*; African Union: Addis Ababa, Ethiopia, 2015.
31. Smith, S. A critical review of the bioavailability and impacts of heavy metals in municipal solid waste composts compared to sewage sludge. *Environ. Int.* **2009**, *35*, 142–156. [CrossRef] [PubMed]
32. WMRA. *Egyptian Code of Design Principal and Implementation Conditions for Municipal Solid Waste Management Systems*; WMRA: Harrisonburg, VA, USA, 2019.
33. Youm7. Egypt's New Waste Management System. 2021. Available online: www.youm7.com/story/2021/2/13/%D9%85%D8%B5%D8%B1-%D8%AA%D8%AA%D8%AC%D9%85%D9%84-%D9%85%D9%86%D8%B8%D9%88%D9%85%D8%A9-%D8%A7%D9%84%D9%85%D8%AE%D9%84%D9%81%D8%A7%D8%AA-%D8%A7%D9%84%D8%B5%D9%84%D8%A8%D8%A9-%D8%A7%D9%84%D8%AC%D8%AF%D9%8A%D8%AF%D8%A9-%D8%AA%D8%B3%D9%8A%D8%B1-%D8%B9%D9%84%D9%89-%D9%82%D8%AF%D9%85-%D9%88%D8%B3%D8%A7%D9%82/5204062 (accessed on 13 September 2021).
34. MoEnv—Ministy of Environment. *Waste Sector Green Growth National Action Plan 2021-2025*; The Hashemite Kingdom of Jordan: Amman, Jordan, 2020.
35. Abu-Ashour, J.; Abu Qdais, H.; Al-Widyan, M. Estimation of animal and olive 219 solid wastes in Jordan and their potential as a supplementary energy source: An overview. *Renew. Sustain. Energy Rev.* **2020**, *14*, 2227–2231. [CrossRef]
36. Abu-Qdais, H.A.; Al-Ghazo, M.A.; Al-Ghazo, E.M. Statistical analysis and characteristics of hospital medical waste under novel Coronavirus outbreak. *Glob. J. Environ. Sci. Manag.* **2020**, *6*, 1–10. [CrossRef]

37. El Jalil, M.H.; Elkrauni, H.; Khamar, M.; Bouyahya, A.; Elhamri, H.; Cherkaoui, E.; Nounah, A. Physicochemical characterization of leachates produced in the Rabat-Salé-Kénitra Region landfill technical center and monitoring of their treatment by aeration and reverse osmosis. *E3S Web Conf.* **2020**, *150*, 02013. [CrossRef]
38. World Bank. *What a Waste 2.0. A Global Snapshot of Solid Waste Management to 2050*; World Bank: Washington, DC, USA, 2018. [CrossRef]
39. The Climate Chance Observatory Team. *Moroccan Society's Uneven Response to the Proliferation of Waste. Case Study Morocco*; The Climate Chance Observatory Team: Paris, France, 2019.
40. Hemidat, S.; Oelgemöller, D.; Nassour, A.; Nelles, M. Evaluation of key indicators of waste collection via GIS techniques as a planning and control tool for route optimization. *Waste Biomass Valorization* **2017**, *8*, 1533–1554. [CrossRef]
41. Cooperation GIS. *Report on the Solid Waste Management in Algeria*; Cooperation GIS, SWEEP-Net: Algiers, Algeria, 2014.
42. MPED. *Egypt Vision 2030*; MPED: Cairo, Egypt, 2016.
43. UNEP. *Reducing Plastic Bag Consumption in Egypt: SwitchMed in Egypt–Factsheet*; UNEP: Nairobi, Kenya, 2020.
44. MOLA; CVDB. *Ministry of Local Affairs Report, Jordan. Development of a National Strategy to Improve the Municipal Solid Waste Management Sector in the Hashemite Kingdom of Jordan*; Report on the Implementation Arrangements for the Recommended MSWM Strategy in the Hashemite Kingdom of Jordan (3rd Draft Report); Consultants: LDK, MOSTAQBAL; Ministry of Local Affairs: Amman, Jordan, 2014.

Article

Practical Challenges and Opportunities for Marine Plastic Litter Reduction in Manila: A Structural Equation Modeling

Guilberto Borongan [1,2,*] and Anchana NaRanong [1]

1 Graduate School of Public Administration, National Institute of Development Administration, Bangkok 10240, Thailand; anchana8@gmail.com
2 Regional Resource Centre for Asia and the Pacific, Asian Institute of Technology, Klong Luang 12120, Thailand
* Correspondence: guilberto@ait.ac.th

Abstract: Land-based plastic pollution has increased to the level of an epidemic due to improper plastic waste management, attributed to plastic waste flux into the marine environment. The extant marine plastic litter (MPL) literature focuses primarily on the monitoring and assessment of the problem, but it fails to acknowledge the link between the challenges and opportunities for MPL reduction. The study aimed to examine the practical challenges and opportunities influencing the reduction of marine plastic litter in Manila in the Philippines. Data collected through an online survey from 426 barangays were analyzed using structural equation modeling (SEM) and were then validated using interviews and focused group discussions. Good internal consistency (0.917) and convergent and discriminant validity were achieved. The empirical study has established structural model fit measures of RMSEA (0.021), SRMR (0.015), CFI (0.999), and TLI (0.994), with a good parsimonious fit of the chi-square/degrees of freedom ratio of 1.190. The findings revealed that environmental governance regarding waste management policies and guidelines, COVID-19 regulations for waste management, community participation, and socio-economic activities have positively affected marine plastic litter leakage and solution measures. Environmental governance significantly and partially mediates the effects of, e.g., COVID-19-related waste and socio-economic activities on MPL leakage. However, there is no relationship between the waste management infrastructure and environmental governance. The findings shed light on how to enhance environmental governance to reduce marine plastic litter and address Manila's practical challenges.

Keywords: environmental governance; marine plastic litter; structural equation model analysis; solid waste management

1. Introduction

Without coordinated intervention, the annual flow of plastics into the ocean is expected to nearly triple by 2040, from 11 million tons today to 29 million metric tons, globally [1]. Marine plastic pollution presents significant risks to the marine environment and has been attributed to land-based plastic leakage from improper plastic waste management systems. Five ASEAN countries in 2016, including Indonesia (4.28 metric tons (Mt)), the Philippines (1.01 Mt), Vietnam (0.57 Mt), Thailand (1.16 Mt), and Malaysia (0.33 Mt) are among the top ten countries with this problem in the world, accounting for 28 percent of the land-based marine plastic litter (MPL) that could end up in the ocean [2,3]. As a result, marine plastic litter issues should be addressed in a holistic, land-to-sea approach. According to reports, most ASEAN countries have developed a roadmap for reducing marine plastic litter in accordance with the ASEAN Framework of Action on Marine Debris. The country's MPL roadmap must be enacted and translated into action at the local and city level.

The Philippines is no exception, ranking third in the world in terms of ocean plastic waste leakage, with 0.28–0.75 million Mt per year (after China and Indonesia, which are first and second, respectively) [4]. Manila Bay, situated in Manila, is where the challenges

Citation: Borongan, G.; NaRanong, A. Practical Challenges and Opportunities for Marine Plastic Litter Reduction in Manila: A Structural Equation Modeling. *Sustainability* **2022**, *14*, 6128. https://doi.org/10.3390/su14106128

Academic Editors: Hani Abu-Qdais, Anna Kurbatova and Caterina Picuno

Received: 26 March 2022
Accepted: 11 May 2022
Published: 18 May 2022

Publisher's Note: MDPI stays neutral with regard to jurisdictional claims in published maps and institutional affiliations.

Copyright: © 2022 by the authors. Licensee MDPI, Basel, Switzerland. This article is an open access article distributed under the terms and conditions of the Creative Commons Attribution (CC BY) license (https://creativecommons.org/licenses/by/4.0/).

related to plastic pollution are of great importance nationally and, thus, make headlines globally, as plastic waste that is not properly managed has increased the economic and environmental effects of marine plastics.

Many notable scholars have argued that plastic leakages caused by land-based mismanagement are related to socio-economic activities along the value chain—plastic use and production and domestic and retail consumption, as well as plastic disposal (end-of-life) via unproductive waste management services [3,4]. Willis et al. (2018) [5] evaluated the most effective policies and strategies for reducing plastic pollution and provided a variety of evidence bases for decision-making in addressing the challenges of marine plastic litter and its pressures on the environment, the economy, and society. However, other scholars argue that designing and implementing legitimate, effective, and efficient actions need to be built on a complete understanding of the context of local governance at the city level [6,7]. Effective environmental governance at various levels, e.g., community and/or city level, is, thus, crucial for identifying solutions to the above-mentioned challenges and potential opportunities. The practices, guidelines, policies, and institutions that shape human interaction with the environment are referred to as environmental governance [8] (UNEP Factsheet Series, n.d.). At the global governance forum on 2 March 2022, at the UN Environment Assembly in Nairobi, 175 countries endorsed a historic resolution to end plastic pollution and forge an international, legally binding agreement by the end of 2024. The resolution, entitled *"End Plastic Pollution: Toward an internationally legally binding instrument"*, stipulated, among other provisions, "an affirmation of an urgent need to strengthen global governance to take immediate actions towards long-term elimination of plastic pollution" [9] (UNEP, 2022). With this global and binding agreement, cities like Manila could properly enforce measures to end this plastic pollution—which needs the political will of the administrators and the participation of the relevant constituents. In addition, action against marine plastic pollution has been linked to the UN's Sustainable Development Goals (SDGs), such as SDG 6 (regarding clean water and sanitation); SDG 11 ("Make Cities and Human Settlements Inclusive, Safe, Resilient, and Sustainable"); SDG 12 ("Ensure Sustainable Consumption and Production Patterns"); and SDG 14 ("Conserve and Sustainably Use the Oceans, Seas, and Marine Resources for Sustainable Development"). Similarly, shifting to more sustainable production and consumption practices, which are also promoted by the SDGs, has been suggested as a solution to marine litter [7].

Whereas past studies have focused on the context of plastic waste pollution for upstream research activities, such as the circular economy and waste management, the research framework, and coordination, very little research on environmental governance, e.g., laws, administrative measures and action plans, guidelines, and current standards [10]. Yang, Y. et al. (2021) [11] proposed countermeasures, including environmental governance, to accelerate China's abatement of marine plastic waste. Moreover, their research highlighted the importance of establishing and implementing an accountable and responsible marine plastic waste governance system. To the best of the authors' knowledge, there has been no empirical research on the relationships between the challenges and opportunities among factors affecting the reduction of marine plastic litter. Although there are studies on the challenges and opportunities for MPL reduction, specifically, the clean-up campaign drive, waste separation, and recycling, most of them are descriptive and qualitative. It is necessary to conduct an empirical investigation into the relationships between factors (challenges and opportunities) affecting the reduction of marine plastic litter. Furthermore, because of the current COVID-19 pandemic there has been an increase in the use of plastics and their subsequent disposal, in the form of personal protective equipment such as face masks, single-use disposable food containers from food delivery services, and e-commerce from online package shipping. Several researchers emphasized how the disruption caused by COVID-19 can be a catalyst for change in global plastic waste management practices in the short and long term [12,13], as they proposed to mitigate the likely impacts of the COVID-19 pandemic on waste management systems.

It is evident that the financial, human, and environmental costs of poor waste management are rising. People living within or near disposal sites, for example, have insufficient access to clean water; tourism development, which is one of the key drivers of economic growth and investment, is linked to the coastal urban ecosystem, which is under threat from plastic found along ocean shorelines and on beaches. The study adds value in contributing to the existing literature for mitigating the challenges of marine plastic litter in coastal cities. Considering the challenges and opportunities for marine plastic pollution reduction, this study aims to examine the practical challenges and opportunities influencing the reduction of marine plastic litter in Manila in the Philippines.

2. Literature Review and Hypothesis Development
2.1. Waste Infrastructure and Environmental Governance

In terms of the changes of administration of most local governments, policy should include de-risking waste infrastructure investment to encourage private sectors to engage. Complex infrastructural systems comprising technologies, regulations, public services, and user practices are required to address urban waste, yet there were no links between waste infrastructure and governance in a previous empirical study [14]. Soltani et al., (2017) [15] report that the ideal waste facility and technology option will fit in with municipality/city objectives, as well as help to save resources in terms of the environmental and financial resources of the community. It was suggested that central and local governments will need to formulate policies to encourage private sectors to invest in their waste infrastructure or technology to effectively reduce MPL [16]. However, Kenisha, G. et al., (2017) [17] conceded that in terms of MSW facilities and infrastructures, public acceptance is vital to ensuring the effectiveness of waste strategies; the authorities in the municipality and city need to seek a practical approach in engaging communities and stakeholders in the decision-making process. Nevertheless, there is a need for waste infrastructure with local governments in public spaces for the effective and efficient implementation of SWM policy, e.g., waste bins along the shoreline or beaches; this will be part of the environmental governance actions at the city and barangay levels. Building on the literature, we propose the following hypothesis:

Hypothesis 1 (H1): *The available waste infrastructure (WInfras) will positively affect environmental governance (EG).*

2.2. Environmental Governance Related to COVID-19 Waste

The contemporary literature suggests that environmental governance on some specific types of municipal waste has visibly increased during the COVID-19 pandemic, when communities and cities experienced the highest generation of waste for plastic packaging and food waste. While this situation has put additional pressure on waste management systems, it has proven useful in terms of insights for city administrations and municipal utilities on consumption patterns during emergency situations. Moreover, Benson, Fred-Ahmadu, et al. (2021) [18], and [19] Shiong et al. (2021) provided insights on plastic waste management status, especially PPE, during the COVID-19 pandemic—which emphasized the sudden spike of medical waste that has had a large impact on plastic waste management. Conversely, Benson, Bassey, et al. (2021) [20] suggested that designated waste separation facilities be provided at marked points in different areas to collect used PPEs—as part of the urgent need for effectively handling COVID-19-related healthcare waste. We observe from the above-mentioned studies that environmental governance is associated with emerging issues on COVID-19 waste management; however, there is no empirical research on the correlations between this challenge and MPL reduction. Based on these earlier studies, we propose the following hypotheses:

Hypothesis 2 (H2): *Environmental governance regarding the "management of COVID-19-related healthcare waste" (EGcv) will positively affect existing environmental governance (EG).*

Hypothesis 3 (H3): *Environmental governance regarding the "management of COVID-19-related healthcare waste" (EGcv) will positively affect marine plastic litter solution measures (MPLr).*

2.3. Community Participation

Regarding the community participation factor, GESAMP 2019 [21] reports an overview of key value-chain stages corresponding to stakeholders/interest groups, and the consequences of environmental plastics connected to each value stream and level. Conversely, Deutsche Gesellschaft für Internationale Zusammenarbeit (GIZ) [22] reported the fundamentals of extended producer responsibility (EPR) for packaging and the many roles that stakeholders might play in the plastic packaging value chain. The study goes through numerous possibilities for allocating duties, as well as the actions that must be followed to reach an agreement and lay the groundwork for the implementation of an EPR system. Producers, retailers, distributors, consumers, and local and central governments are key stakeholders in the plastic packaging value chain, e.g., most local governments are responsible for the collection of plastic packaging [23]. Wilson et al., 2015 [24] demonstrate that the SWM system is made up of two intersecting "triangles", one for physical variables such as collection, recycling, and disposal, and the other for governance factors such as inclusion, financial sustainability, sound institutions, and proactive policies. Experience confirms the utility of indicators in allowing comprehensive performance measurement and comparison of both "hard" physical components and "soft" governance aspects, and in prioritizing the "next steps" in developing a city's solid waste management system, by identifying both local strengths that can be built on and weak points that must be addressed [5,24–26]. The private sector, a powerful actor, is able to negotiate and adjust regulations for its own benefit [27]. The waste policy also has a contemplative and encompassing responsibility to the private sector in order that manufacturers, distributors, and importers who are contributing market products, which are eventually turned into waste, should also be responsible for contributing to the recycling or disposal cost. Implementation of the planned interventions needs commitment not only from the government but also from the public and private sectors [28]. Based on the above studies, we proposed the following hypotheses:

Hypothesis 4 (H4): *Community participation (CParty) is positively related to environmental governance (EG).*

Hypothesis 5 (H5): *Community participation (CParty) is positively related to marine plastic litter leakage (MPLe).*

2.4. Socio-Economic Activities Related to MPL Pollution

Large volumes of plastic litter are transported to the sea or ocean through rivers, adding to the serious environmental, economic, and social issues of marine litter contamination [29]. The study by Adam et al. (2021) [30] emphasizes the effect of residents' attitudes and behaviors regarding single-use plastics in Ghana's coastal cities of Accra and the Cape Coast. The significance of their results for reducing marine single-use plastic pollution includes policies and programs, particularly those that are behavioral in character and are built on the idea that the public has a variety of emotions and behaviors. Socioeconomic activities differ depending on "socio-demographic factors" (e.g., gender), political orientation, marine contact factors (e.g., maritime occupations and participation in coastal recreation activities). There is a great deal of evidence that plastic has a harmful influence on marine wildlife and ecosystems [31]. Moreover, marine plastic litter is having an increasing influence on the environment, human health, and economies in the South Pacific [32]. However, the study findings demonstrate how general attitudes about climate change can influence both climate policy support and personal climate mitigation behavior in both direct and indirect ways, giving crucial insights useful for research and policy-making. From the previous studies discussed herein, we propose the following hypotheses:

Hypothesis 6 (H6): *Socio-economic activities (SEas) are positively associated with environmental governance (EG).*

Hypothesis 7 (H7): *Socio-economic activities (SEas) are positively associated with marine plastic litter solution measures (MPLr).*

Hypothesis 8 (H8): *Socio-economic activities (SEas) are positively associated with marine plastic litter leakage (MPLe).*

2.5. Manila Public Behavior Related to MPL Pollution

The types and origins of marine plastic litter vary greatly, ranging from direct losses from recreational and commercial ships and vessels in seas and rivers to indirect losses produced by land-based sources in conjunction with the plastic value chain [29]. Several distinguished scholars have argued that plastic leakage caused by land-based mismanagement is related to plastic use and production and to domestic and retail consumption, as well as plastics disposal (end-of-life) via unproductive waste management services [3,4]. Asia has driven the growth in plastic production over recent decades. It is now the leading plastic consumer in this region, with per-capita plastic use growing at a faster rate than in other regions. The year 2017 saw the global production of 348 million tons of plastic [33] and in the next two decades, the total volume of plastics that will be produced is projected to double [34]. As a result of ocean currents, the leaked ocean plastic waste can potentially travel lengthy distances to other areas and countries—which makes it transboundary in nature. Plastic waste pollution has even started to travel to isolated places, leading to the current challenge and making its prevention globally significant [3]. There is evidence that marine plastic pollution has a substantial economic impact, especially on "fisheries, aquaculture, recreation, and heritage values". Moreover, marine litter has a negative impact on Small Island developing states, owing to their limited waste disposal infrastructure. According to researchers, marine plastic waste has a spillover effect on aquatic marine life, posing severe health concerns for aquatic marine life and maybe even to humans if they consume it. However, the problem is exacerbated by population increase, economic industrialization, a lack of tools to improve collection rates, and the existence of landfills in metropolitan areas, which has a detrimental influence on public health [32,35–37]. Based on the previous studies, we propose the following hypotheses:

Hypothesis 9 (H9): *Manila public litter behavior or the MPL problem (SE) is positively associated with marine plastic litter leakage (MPLe).*

Hypothesis 10 (H10): *Manila public litter behavior or the MPL problem (SE) is positively associated with marine plastic litter solution measures (MPLr).*

2.6. Environmental Governance

Environmental governance brings forth the underlying institutional theory [38], which tends to be associated with the institutional environment, such as the political, cultural, and social processes. While environmental governance on SWM or marine plastic litter exists, waste and marine plastic pollution is governed by actors beyond formal government; however, it is not clear from the policy statements and documents how the various actors in the different spheres of governance interact. An amalgamation of institutional theory and resource dependence theory underpins this to enhance the strength of the theories utilized for enhancing environmental governance in the local context. Oliver's contribution reveals how institutional and resource-dependence theories can be combined to discover a variety of strategic and tactical responses to the institutional environment and other elements [37]. While Oti-Sarpong, K., et al. (2022) [39] used institutional theory to examine the factors driving the increased use of offsite manufacturing to construct new housing in selected countries, their findings highlight the need for more institutional theory research into off-site manufacturing to better understand path dependence. We used these theories to investigate the relationship between environmental governance and resource availability in barangays and in the city of Manila for tackling SWM and marine plastic litter. The mentioned amalgamation of theories averred an integrated solution for the abatement of MPL. It is worthy of note that researchers continue to uncover vital approaches to the momentum of the diffusion of knowledge to aid decision-makers in addressing marine plastic pollution challenges, in order that they would, in turn, be able to assist the community and city. Previous studies have supported the evolution of environmental governance grounded on historical screening, on a level and integration that cannot be reflected upon without consideration of the temporal aspect [40]. However, Whiteman, A., Smith, P., and Wilson (2001) [41] explicitly elaborated on environmental governance to assess the performance of the three main aspects of governance, such as inclusivity of stakeholders, financial sustainability, and sound institutions and proactive policies. On the other hand, Glasbergen (1998) [42] "identifies and describes five main models of environmental governance, these include: regulatory, market regulation, civil society, co-operative, contextual control, and self-regulation". The works of Willis et al. (2018) [5] suggest that the combined solution of the applied model to reduce waste volumes includes litter prevention, recycling, and illegal dumping—which result in the significant reduction of plastic waste in the local government's coastal areas [5]. Furthermore, previous research suggests that municipalities or cities that invest and/or spend on waste management, as well as on a fund for coastal initiatives, have reduced the waste burden in their coastal areas. Other scholars, such as Breukelman et al. (2019) [43], acknowledge that more research is needed using diagnostic analysis regarding the failure of SWM services in the cities of developing countries to better enable interventions to address impacts such as marine plastic litter. "The success of a city's SWM system can be used as a proxy indication of excellent governance," according to one scholar. Most of the existing literature has stressed that the key practical challenge in SWM is a lack of data and data consistency when comparing cities. Moreover, the existing literature calls for indicator sets for integrated sustainable waste management (ISWM), for benchmarking SWM effectiveness in developed and developing cities, particularly for monitoring applications [23,44]. Based on the previous peer-reviewed research, we proposed the following hypotheses:

We postulate that the impacts of EG on COVID-19 waste management (EGcv) and socio-economic activities (SEas) on MPL leakage (MPLe) are mediated by environmental governance (EG), as follows.

2.6.1. Mediated Effects

Hypothesis 11 (H11): *Environmental governance (EG) mediates the relationship between EG on COVID-19 waste management (EGcv) and MPL leakage (MPLe).*

Hypothesis 12 (H12): *Environmental governance (EG) mediates the relationship between socio-economic activities, the marine litter problem (SEas), and MPL leakage (MPLe).*

2.6.2. Direct Effects

Hypothesis 13 (H13): *Environmental governance, in terms of strategies, guidelines, and implementation procedures (EG), will positively affect marine plastic litter solution measures (MPLr).*

Hypothesis 14 (H14): *Environmental governance, in terms of strategies, guidelines, and implementation procedures (EG), will positively affect marine plastic litter leakage (MPLe).*

2.7. Marine Plastic Pollution (MPL) Solution Measures

In terms of solution measures factor, Wu, (2020) [45] argued that rapid urbanization and industrialization cause a great deal of industrial and municipal waste, triggering environmental and human health issues. The impact leakage pathway framework explained by Alpizar et al. (2020) [29] identifies important policies for institutional aspects. However, scholars argued that legislation has improved waste-related practices in businesses, as it was observed that waste legislation is fragmented, and taxation incoherent [46–48]. It was contended that the abatement of marine litter generation will work directly at the local source (including sweeping, collection, single-use bag bans, and other activities) through financing and contracting out solid waste management systems. This stage should include educational campaigns, efforts toward litter reduction, cleanup activities, and law enforcement mechanisms [49]. In fact, the success of devices such as deposit–refunds is largely determined by consumers' willingness to compensate for waste-related environmental damage [50]. Binetti et al. (2020) [32] suggested measures to minimize single-use plastic, enhance collection, reuse, and recycling, as well as creating public awareness campaigns, which might considerably reduce marine litter.

2.8. MPL Leakage

GIZ explicitly described cases in most of the ASEAN developing countries where most of the "uncollected plastic waste is either burned or disposed of into waterways, thus leading to the partial leakage of such plastic waste into rivers". Marine litter comes from sources on land (UNEP, 2016) [51] and its localized abundance is linked to urbanization and the levels of waste management infrastructure, as well as to recreational activity [4,10]. People using the riverside as a recreational area, residents without access to adequate waste infrastructure, people illegally disposing of litter, wastewater treatment plant outlets or sewage overflow, and the plastic-producing or plastic-processing industry are all sources of anthropogenic litter at riversides. However, several studies have indicated an increase in waste volumes downstream of bigger urban areas, and many of these sources are linked to densely populated places (i.e., cities or urban spaces) [3,52–56].

Multigroup Effects

Hypothesis 15 (H15): *The positive relationship between socio-economic activities (SEas—plastic pollution problem) and MPL leakage is stronger for females.*

2.9. Theoretical Framework of MPL Reduction

Based on the literature, we established the following model, showing the practical challenges that influence the endogenous variables that marine plastic litter solution measures (MPLr) and marine plastic litter leakage (MPLe) may impact, with exogenous variables; this is anchored in institutional theory and resource dependence theory. Practical challenges include waste infrastructure, community participation, physical socio-economic activities, and, with environmental governance as a mediating variable, the need to address the mentioned specific research questions in the case of Manila in the Philippines (shown in Figure 1). The theoretical framework depicted in this section was used for the SEM analysis.

2.10. Structural Equation Modeling

Structural equation modeling (SEM) establishes the link between the measurement model and the structural model, based on the assumptions supported by theory. Factor analysis and linear regression are combined in this method of SEM [57]. The difference between regression and SEM decision-making approaches is that regression models are additive, whereas structural equation models are relational. Structural equation modeling also investigates the direct and indirect effects of mediators in the relationship between the independent and dependent variables, in order to support the acceptance or rejection of a hypothesis.

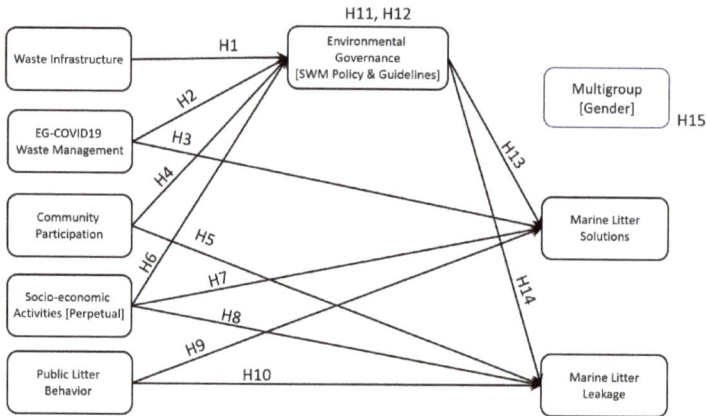

Figure 1. The theoretical framework of MPL reduction.

In line with the developed theoretical framework of marine plastic litter reduction (shown in Figure 1) and the existing literature presented in this paper, the model provides a framework to address the research questions: (1) Can the practical challenges of waste infrastructure, environmental governance, community participation, socio-economic activities and public litter behavior empirically influence the reduction of marine plastic litter in Manila? (2) What are the drivers of marine litter within Manila and the Philippines context? (3) What are the long-term opportunities for the reduction of marine plastic litter in Manila?

3. Materials and Methods

3.1. Manila Demography and SWM

A case study of Manila is the focus of this paper. Manila is the Philippines' capital and the country's most densely populated city. In 2015, Manila had a population of 1.78 million people. Manila is divided into 896 urban barangays, the Philippines' smallest unit of local government. Each barangay has its own councilors and chairperson. All of Manila's barangays are divided into 100 zones for administrative purposes, which are then divided into 16 administrative districts. There is no local government in these zones and districts. Trade and commerce are the city's pillars. North Manila, which is located on the upper portion of the Pasig River, and South Manila, which is located on the lower portion of the river, have distinct characteristics. Manila Bay and Laguna de Bay are connected by the Pasig River, which is about 25 km long. Over 2000 factories and 70,000 families live in makeshift shelters along the river's banks [58]. These bodies of water may have the potential for MPL littering and leakage to the ocean if environmental governance is not effectively acted upon.

Manila's waste generation accounts for 69.87% of waste from residential sources, while 25.73% is commercial, 1.19% is institutional, 0.19% is industrial, 1.56% is from markets and 1.45% is from street-sweeping (2015 baseline data, [59] Manila DPS, 2020). For waste composition from household and non-household sources, kitchen waste was 39.73%; this was followed by plastic waste at 17.75%, and 9.04% of paper waste, among other wastes [59]. This indicates that the plastic waste stream is quite high in a highly urbanized city like Manila. In the Philippine MSW definition, waste components are composed of biodegradable, recyclable, special waste, and residual waste, wherein the Manila waste composition is at 50%, 32%, 13%, and 5%, respectively. In the baseline data in 2015, Manila has 0.607 kg/capita/day of waste generation, with an annual waste generation of 376,008.40 Mt per year [59]. Of this, 67.69% of waste was sent to landfills, while 32.31% was diverted waste. Through a private contract with the private sector, the city of Manila has a 100% coverage clean-all solid waste collection and disposal system.

The Department of Public Services is responsible for the city's solid waste management and environmental sanitation, under the direct supervision of the department head. The organizational structure of the DPS comprises four (4) divisions and six (6) district offices, to enhance the efficient dispensation of DPS activities. All municipal rules and ordinances related to solid waste management and other environmental problems in the city of Manila are implemented and enforced by the DPS.

Manila's governance structure is governed by the City Solid Waste Management Board (CSWB), which was established in 2000 in accordance with the Philippine Ecological Solid Waste Management Act. The city has a 10-year SWM plan that has been approved (2015–2024). The City Mayor (Chairman), City Administrator (Vice-Chairman), and Head of DPS, among other relevant constituents, must actively participate in the CSWB's specialized responsibilities and functions [59]. The Department of Public Services is in charge of enacting and enforcing all municipal rules and ordinances relating to solid waste management and other environmental issues in Manila (DPS). Over the years following its approval, the city legislators created and approved the necessary city ordinances, in accordance with the Ecological Solid Waste Management Act of 2000 (R.A. 9003). Executive Orders issued by the mayor (past and present) serve as a backbone for enjoining the involvement and commitment of the city and barangay authorities to implement, enforce, and support all local legislation related to SWM and environmental preservation.

3.2. Framework Development, Data Collection, and Analysis

The conceptual framework process diagram in Figure 2 depicts the process flow of the empirical quantitative SEM method and qualitative approach used in this study, which began with the literature review phase. Data were gathered using a Google form for an online survey. The study's survey items were adapted from previous studies (as described in Table 1) [3–5,29,30,32,34,35,60–64] related to the reduction of solid waste management and marine plastic litter from land-based sources. Additional survey items were developed and adapted from the Philippine interim COVID-19 waste management guidelines. Structural equation modeling was used to analyze the collected data. A pretest survey was also conducted prior to the formal online survey distribution. Using the MPL reduction framework, we used exploratory factor analysis, confirmatory factor analysis, and structural equation modeling (SEM) analysis. The findings and suggested recommendations for the study were quantitatively analyzed using an empirical SEM method, validated through interviews and focused group discussions (FGD) in Manila with the relevant constituents/experts in SWM in the Philippines.

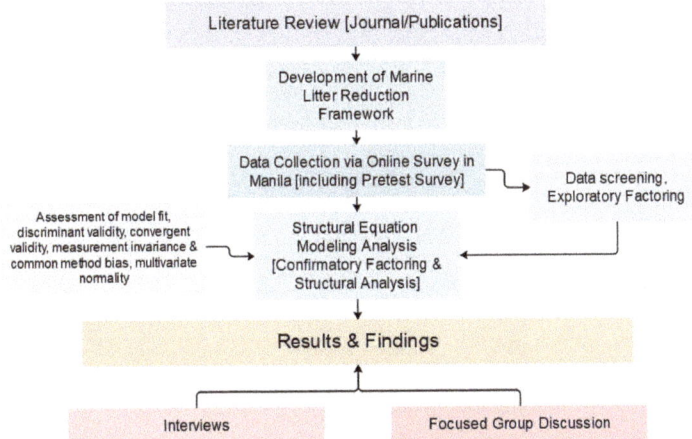

Figure 2. The study's conceptual framework process diagram.

Table 1. Model Latent Factor with corresponding manifest indicators, descriptions, and methodological backing.

Latent Factor (Nomenclature)	Manifest Indicator	Description	Methodological Backing
Environmental governance (EG: Policy and guidelines on SWM/MPL)	EGa1	Clear guidelines and strategy for MPL, SWM	Willis, K., et. al. (2018); Wilson, D. C., et. al. (2015); [5,24], F. Alpizar, F. Carlsson (2020). [29] Glasbergen (1998) [42] Plummer R., et al. (2013) [65] Wilson, et al., (2015) [24] Lyons et al., 2020 [10] Allan Paul Krelling, et al., (2021) [63]
	EGa2	Effective mechanisms in place for waste facility	
	EGa3	Openness, transparency, and accountability of bid processes in SWM	
	EGa4	Institutional arrangements for SWM	
	EGa5	Institutions SWM budget	
	EGa3F1	SWM accounts reflect accurately the full costs of providing the service, the relative costs of the different activities within SWM	
	EGa3F2	Annual budget adequate to cover the full costs of providing the SWM service	
	EGa3F3	Percentage of the total number of households both using and paying for 'primary waste collection services	
	EGa3F4	Practices or procedures in place to support charges/fees	
	EGaI1	Public involvement at appropriate stages of the SWM decision-making, planning and implementation process	
Environmental governance (EGcv: COVID-19-related Healthcare Waste Management)	EGCV1	Awareness of the "Interim guidelines on the management of COVID-19-related healthcare waste"	Interim COVID-19 Waste Guidelines (Adopted and modified from DENR, 2021) [66] Benson, N.U. et al., 2021. [18] Benson, Bassey, et al. (2021) [20] Kuan Shiong Khoo, et al., 2021 [19]
	EGCV2	Use of the COVID-19 related waste management plan template	
	EGCV3	Proper handling and management of all COVID-19-related health care waste	
	EGCV4	Manage and contract waste service providers ("waste separation and collection, transport, treatment, and disposal") in accordance with the adopted LGU COVID 19 plans	
	EGCV5	Orientations on COVID 19 proper waste management	
Waste Infrastructure (WInfras): Existing Technical/Waste Facilities	T1	Landfill is near waterways/rivers	Wilson, D. C., et. al. (2015) [24] Willis, K., et. al. (2018) [5]
	T2	Existing MRF	
	T3	Artificial or special catching barriers (screen traps) to stop waste entering the sea	
	T4	Boats cleaning the waterways, estuaries, rivers, or sea exist	
	T5	Diversion programs, e.g., recycling to abate the marine plastic litter	
	WI1	Waste of collection points/transfer stations	Willis, K. et al. (2018) [5] Whiteman, A., Smith, P. and Wilson, D.C., 2001. [41] B.P. Lyons, et al., (2020) [10]
	WI2	Effectiveness of street cleaning	
	WI3	Efficiency and effectiveness of waste transport, e.g., garbage trucks	
Community Participation (CParty): 3Rs and Circular Economy Roles and responsibility to enhance packaging waste management (3Rs—reduce, reuse, recycle) in the Philippines	CP1	Consumers	Cai et al., 2021 [44] PREVENT Waste Alliance, 2020 [23] Oke et al., 2020 [25] Morten W. Ryberg, Alexis Laurent, 2018 [26] Wilson, D. C., et. al. (2015) [24] Willis, K., et. al. (2018) [5] (GIZ, 2018) [22]
	CP2	Plastic producing industry (raw material)	
	CP3	Filters and importers	
	CP4	Retailers of plastic items (e.g., supermarkets)	
	CP5	Government	
	CP6	Local authorities	
	CP7	Associations/NGOs	
	CP8	Scientific institutions/academia	

Table 1. Cont.

Latent Factor (Nomenclature)	Manifest Indicator	Description	Methodological Backing
Marine Plastic Litter Reduction Solution Measures (MPLr)	MPLsm1	A law to introduce extended producer responsibility (EPR)	(GIZ, 2018) [22] PREVENT Waste Alliance, GIZ, 2021. [23] Willis, K. et al. (2018) [5] Wilson, D. C., et. al. (2015) [24] UNEP (2016) [51]
	MPLsm2	Establishing deposit systems for plastic bottles	
	MPLsm3	Establishing material recycling facilities	
	MPLsm4	Enhancing waste collection coverage in barangays	
	MPLsm5	Enhancing waste separation at all households and establishments	
	MPLsm6	Co-processing of plastics in cement plants	
	MPLsm7	Construction of incineration plants	
	MPLsm8	Opening landfills	
	MPLsm9	Awareness-raising campaigns	
	MPLsm10	Conduct (beach/coastal) clean-ups	
	MPLsm11	Banning of single-use plastic products	
	MPLsm12	Producing packaging made from bioplastics	
	MPLsm13	Producing packaging made from alternative materials	
	MPLsm14	Making plastic packaging reusable and recyclable	
	MPLsm15	Introducing (comprehensive) disposal fees	
	MPLsm16	Introducing fines for littering	
Socio-economic activities (SEas)	SEa1	Single-use plastic packaging is an expression of economic prosperity (plastic use from extraction, consumer, post-consumer, disposal, with links to the linear economy)	GIZ, 2018 [22] Jambeck et al., 2015 [4]; L.C.M. Lebreton et al., 2017 [3] Faten Loukil and Lamia Rouched 2012 [50] Sophie M.C. Davison, et al., 2021 [35] Nelms et al., 2020 [36]; Rochman et al., 2016 [64] Binetti et al., 2020 [32]
	SEa3	Threat to the environment, human health, and economic prosperity (industrialization has been associated with an increase in packaging waste)	
	SEa4	Plastic consumption contributes to climate change	
	SEa7	Quantity of plastic pollution in the natural environment is increasing	
MPL Leakage (MPLe: Pathways (routes) contribute to litter in the marine environment)	MPLa1	Litter reaches the sea from rivers, canals, creeks, and estuaries	Alpizar et al., 2020 [29] Gasperi et al., 2014 [55]; Morritt et al., 2014 [56]; Mani et al., 2015 [54]; Lebreton et al., 2017 [3]; Di and Wang, 2018 [53]; Magni et al., 2019 [52] Jambeck et al., 2014 [4]; L.C.M. Lebreton et al., 2017 [3] (GIZ, 2018) [22]
	MPLa2	Litter is blown into the sea from landfills	
	MPLa3	Flooding and sewage overflows	
	MPLa4	Direct release on the coast, e.g., beach users, coastal tourism	
	MPLa5	Direct release in the sea (by fishing, ships, and offshore industries)	
Manila Public Behavior and the MPL Problem (SE: factors in the city contributing to public plastic littering of the environment)	SEb1	Public behavior in terms of littering	Issahaku, Adam, et al. (2021) [61] (GIZ, 2018) [22]
	SEb2	Lack of waste collection and separation	
	SEb3	Lack of adequate waste management infrastructure and facility	
	SEb4	Intense consumption of single-use plastics	
	SEb5	Lack of enforcement of waste disposal directives	
	SEb6	Lack of funding for waste collection	

3.3. Data Analysis Methods for the MPL Reduction Framework

Figure 3 depicts the SEM analytical process at various stages of the analysis, including data screening, EFA, CFA, and the structural analysis of the manifest variables used in the article (both exogenous and endogenous variables). Starting with data screening and descriptive analysis, exploratory factor analysis (EFA), confirmatory factor analysis (CFA), and the path/structural analysis—covariance-based SEM method, the latent variables of socio-economic factors, environmental governance, community participation, waste infrastructure, and other latent variables were plugged into MPL as an abatement latent variable using SEM analysis. An online pre-test survey was conducted with 79 samples of 51 survey items in August 2021, to determine the appropriateness and fit of the use of the survey in the Manila local government unit (LGU) and respective barangays. The reliability statistics of the 79 cases have a Cronbach's alpha value of 0.907. Any additions and revisions of the terms and statements for respondents regarding the usage of terms and ease of understanding of the survey items were performed. The formal online survey questionnaire, using Google Forms, was distributed in the period from 12 to 30 September 2021 to a total of 800 barangays and other Manila respondents. This sample size fits the recommendation of MacCallum, R. C., Browne, M. W., and Sugawara (1996) [67], who suggested a larger case-to-variable ratio. The actual total number of samples collected comprised 456 responses from barangays and Manila DPS respondents. These experiment samples were screened and processed to identify omissions, unengaged responses, and irrelevant items. Finally, 426 out of the 456-questionnaire sample size were processed and analyzed for the study.

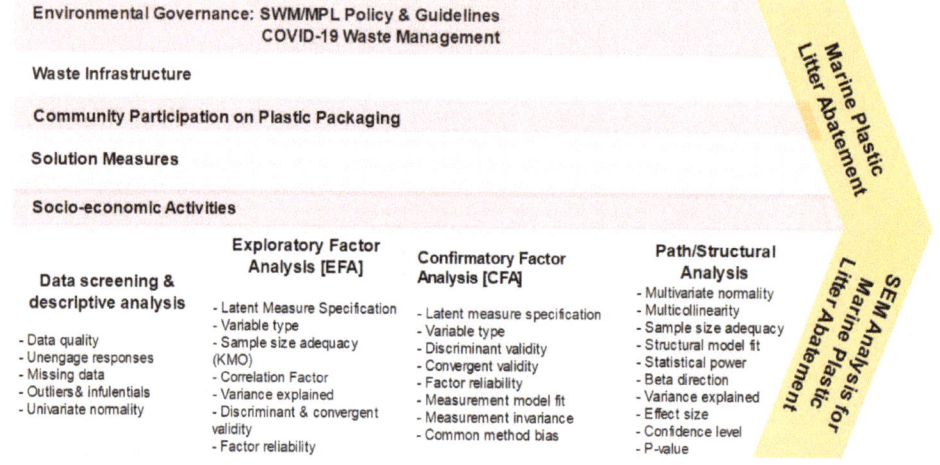

Figure 3. SEM Analysis Process Diagram.

A random sampling method was performed to collect data from the study population, who were asked to rate the 62 indicators based on their level of agreement, using a four-point Likert scale (e.g., anchored at 1 = strongly disagree, 2 = disagree, 3 = agree, and 4 = strongly agree).

An examination of the reliability of the instrument was necessary. For this reason, an attempt was made to check the item-total score correlations, indicating that items with higher correlations are better instruments [68]. At this stage, 62 interval Likert scale indicators and 28 categorical Likert items were utilized for the survey.

The marine plastic litter reduction framework was used to draw out suggested policy and action implications and address the research questions of the study. Moreover, the conceptual framework was validated through interviews and focused group discussions (conducted from 15 November to 10 December 2021) in Manila and with relevant Philippine SWM experts. Target participants for the virtual interview and FGD were the assistant city government head and technical staff from Manila DPS, a representative from the barangays, the NSWMC Officer in Charge, selected SWM contractors from the private sector, and NGOs/Civil Society "Solid Waste Association of the Philippines" (SWAPP) and

Academe members working on SWM and marine plastic litter reduction. The contacts and/or email addresses that were used for the interview and FGD (through Google Meet and the WEBEX platform) came from the author's existing established network in Metro Manila, the endorsement of the Manila Department of Public Services, and the League of Cities in the Philippines, as well as from the Philippine National Solid Waste Management Commission (NSWMC).

4. Results and Discussion

4.1. Demographic Characteristics of the Respondents

The demographic characteristics of the Manila respondents are illustrated in Table S1 (in the Supplementary Materials). Demographic information includes gender (male (35.7%) and female (64.3%)) and age group in years (under 25 years old (28.6%), 25–34 (28.6%), 35–44 (26.8%), 45–54 (12%), and over 55 years old (4%)). Most of the respondents have completed undergraduate studies (52.8%), while respondents completing secondary-level education and graduate studies are 24.2% and 23%, respectively. Most of the respondents who took part the survey had a range of income of PHP 10,001–15,000 (pesos) (40.8%). Of the respondents, 92.7% (395) were affiliated by their work to the barangay units, followed by Manila DPS at 4.7%, with 0.2% of representatives from an association/NGO. Similarly, 62.7% (267) of respondents were barangay secretaries who completed the survey forms on behalf of their respective barangay units, while 24.9% (106) of respondents were barangay chairmen/chairwomen, followed by barangay councilors at 8.9% (38). It is important to note that the Barangay Solid Waste Management Committee is composed of the barangay chairman, councilor, secretary, and barangay technical staff members—in line with the R.A. 9003. The number of years of affiliated work (in barangays and Manila LGU) of most of the respondents was 56.6% (241) with under 5 years, followed by 23.7% (101) with 6–10 years, 9.9% with 11–15 years, 5.2% with 16–20 years, and 4.7% with over 20 years of work. The target population of barangays in Manila is 896 barangays. The target sample size was determined to be 277 samples (barangays), which is substantially more than 200 samples, as determined, and was collected in a distributive pattern in demographic barangays/zones in Manila. The larger number of samples was not only desirable for adequate data collection but was intended to be segmented in separate analyses by barangays/zones demographics while still maintaining a credible sample size. In addition, as in the requirements for SEM in Amos version 28, the sample size is adequate to run the CFA and structural model.

4.2. Descriptive Statistics

A total of 62 socioeconomic, environmental governance and MPL variables were used to gauge the respondents' level of agreement. The majority of the respondents expressed their satisfaction with most of the indicators. Factors of socio-economic perception related to Marine Plastic Pollution (MPL) had a mean (M) range of 2.44–3.64, and a standard deviation (SD) range of 0.494–0.819, with a level of agreement of "strongly agree". MPL plastic leakage attributes (MPLe) have a mean (M) range of 3.19–3.37 and a standard deviation (SD) range of 0.615–0.662, which indicates a level of agreement of "very much". In terms of socio-economic factors, public littering behavior had a mean range of 3.53–3.65 and an SD range of 0.517–0.546, which suggests a level of agreement of "very important". The waste infrastructure (WInfras) was represented by 3 indicators, of which 3 indicators scored lower mean values at 2.41–2.60 and a standard deviation range of 0.901–0.988, suggesting a level of agreement of "low incidence to medium incidence". Community participation regarding plastic packaging has a mean range of 3.48–3.65 and an SD range of 0.543–0.610, indicating a level of agreement of "very much". Environmental governance related to policies and strategies/guidelines (EG) mean range was 2.73–2.89, with an SD of 0.705–0.842, suggesting a level of agreement of "medium compliance". Environmental governance related to COVID-19 waste (EGcv) has a mean range of 3.16–3.26 and an SD range of 0.647–0.693, suggesting a level of agreement of "medium compliance". MPL solution measures (MPLr) provided a mean range of 3.22–3.48 and an SD range of 0.595–0.702, indicating a level of agreement of "extremely likely". Most of the indicators referring to socio-economic factors, MPL, environmental governance, and MPL solution measures reported good mean scores and standard deviations.

4.3. Assessment of Multivariate Normality and Multicollinearity

Before the EFA was performed, tests of multivariate normality and multicollinearity were conducted using SPSS 23. The observed values in the P-P plot are closely parallel to the straight line, indicating that the observed values are similar to what we would expect from a normally distributed dataset [69]. Skewness and kurtosis results do not exceed between +2 and −2 (see Table S2 for the data mean, standard deviation, skewness, and kurtosis). Furthermore, no correlation coefficient value greater than 0.8 was identified for any of the observable indicators in the correlation matrix [70]. Hence, multicollinearity is not a problem with these data.

4.4. Verifiability of Latent Factors

Principal component analysis (PCA) was performed using the ProMax rotation for the 62 manifest indicators in exploratory factor analysis (EFA), to assess the dimensionality of the environmental governance, socio-economic activities perception, and MPL indicators. The total variance explained by the thirteen (13) distinctive factors extracted was 68.02%, with an eigenvalue greater than 1.0 (see Table S2: Factors in MPL reduction with mean, SD, skewness, kurtosis, factor loadings, and Cronbach's alpha in the Supplementary Materials). There were no correlations greater than 0.7, indicating convergent validity and discriminant reliability. The overall Cronbach's alpha value was 0.917 for 62 manifest indicators, which is above the suggested benchmark of 0.6 [71,72]. All the commonalities in this study are above 0.400. There are 84 (3.0%) non-redundant residuals with absolute values greater than 0.05 in the residuals computed between the observed and replicated correlations, indicating that non-redundancy residual measures are not a concern.

4.5. Validity and Reliability Results

The KMO measure of sampling size adequacy value of 0.896 in this study is a great or meritorious degree of common variance [70,73]. The Bartlett test of sphericity tests the null hypothesis that the original correlation matrix is an identity matrix. The correlations between indicators were substantial enough for PCA, according to a statistically significant Bartlett's test of sphericity ($p < 0.05$) [74,75]. For these data values, Bartlett's test is highly significant (χ^2 (20,143.984) = 2628, $p < 0.000$); thus, it is safer and appropriate to proceed with factor analysis and CFA. For validity tests in the exploratory factor analysis, 13 latent factors with 62 indicator variables were extracted with eigenvalues of 13.377 to 1.010 and account for 68.022% of the covariance among the manifest variables (as shown in Table S2: Factors in MPL reduction with mean, SD, skewness, kurtosis, factor loadings, and Cronbach's alpha). Forty-seven (46) manifest variables were retained, while 16 indicator variables were removed, in conformity with the assumptions.

The nine (9) latent factors retained were MPL solution measures (13 manifest items), SE attributes (6 manifest items), MPL leakage (5 manifest items), community participation (5 manifest items), environmental governance: SWM policies (5 manifest items), EG on COVID-19 waste management (4 manifest items), socio-economic activities A (4 manifest items), waste infrastructure (3 manifest items) and socio-economic activities B (2 manifest items). From the extracted latent factors, manifest variables in environmental governance in the case of Manila—SWM policies, financial resources, and community participation—were combined based on the PCA (see Table S2: Factors in MPL reduction with mean, SD, skewness, kurtosis, factor loadings, and Cronbach's alpha). Overall, Cronbach's alpha of nine (9) latent factors for the 426 cases was 0.917. The individual latent factor reliability results have a Cronbach's alpha of MPLr (0.957), SE (0.879), MPLe (0.885), CParty (0.913), EG (0.900), EGcv (0.904), SEas (0.780), WInfras (0.791), and SEc (0.708), respectively. The 9 latent factors have coefficient loadings greater than 0.500 (as illustrated in Table S2: Factors in

MPL reduction with mean, SD, skewness, kurtosis, factor loadings, and Cronbach's alpha values) were utilized in the study. Thus, we have a reliable and valid instrument with very good internal consistency.

4.6. Confirmatory Factor Analysis

To examine the validity of the latent variables, confirmatory factor analysis (CFA) was performed. This "concept relates to the degree to which a scale or collection of measures accurately represents the topic of interest," according to [70,73]. Convergent validity, discriminant validity, and content validity are all requirements of the CFA measurement model. The degree to which two scales of the same issue are correlated is measured by convergent validity [70]. For each construct or latent factor, the convergent validity was checked by computing the average variance extracted (AVE), standardized regression weight (SRW), and composite reliability (CR) [76]. All standard regression weights (factor loadings) in the measurement model were significant at $p < 0.001$, in the range of 0.514–0.871. The minimum level of acceptability for all factor variables was greater than 0.30 [74]. An AVE of 0.5 or above is required to evaluate the measurement model's convergent validity. The square root of AVE, which must be larger than the latent variable correlations, is used to assess the discriminant validity [77]. The discriminant value must be greater than the correlation between the latent variables.

4.6.1. Measurement Model Fit Assessment

Using AMOS version 28, SEM was utilized to assess the measurement model fit through confirmatory factor analysis (CFA) of the study on enhancing environmental governance for MPL reduction in Manila in the Philippines. The model fit measures were unconstrained, and the chi-square goodness-of-fit test was significant [78], with $\chi^2/df = 1.507$, and p-value < 0.001. The RMSEA was excellent at 0.035 and the pClose test was not significant (p-value of 1.000), with a GFI of 0.885, for which the ideal is greater than (>) 0.95. The CFI (0.956), IFI (0.957), and TLI (0.952) were greater than (>) 0.95, while SRMR was 0.042 (see Figure 4). MPL reduction measurement model). Even though the values for GFI and AGFI do not exceed the threshold value of > 0.9, "they still met the requirements as suggested by [79,80] where the value is acceptable if above 0.800". All fit indices of the measurement model of the latent factors were above the recommended threshold values [81], as shown in Table 2. Appropriately, the measurement model fit of the 8 latent factors and observed variables was found to be very satisfactory. Even though the p-value of the measurement model is significant ($p < 0.05$), "the model is strongly affected by the large sample size and dependent on the complexity of the current measurement model (sensitive to a complex model and large sample size)". The current measurement model does not employ unnatural constraints to the set of measures. Thus, the overall model fit indices for the measurement model, suggest a very good model fit.

Table 2. Measurement of model fit.

Criterion of Model Fit	Absolute Fit Acceptance	Values of Model Fit	Test Result
RMSEA	<0.08	0.035	Established
SRMR	<0.08	0.042	Established
pClose	>0.05	1.000	Established
GFI	≥0.90	0.885	Acceptable
AGFI	≥0.90	0.868	Acceptable
CFI	≥0.90	0.956	Established
IFI	≥0.90	0.957	Established
TLI	≥0.90	0.952	Established
χ^2/df	<3.00	1.507	Established

Note: RMSEA = root mean square estimation approximation, GFI = goodness-of-fit index, CFI = comparative fit index. TLI = Tucker Lewis index, IFI = incremental fit index, adjusted goodness-of-fit index (AGFI), df = degrees of freedom.

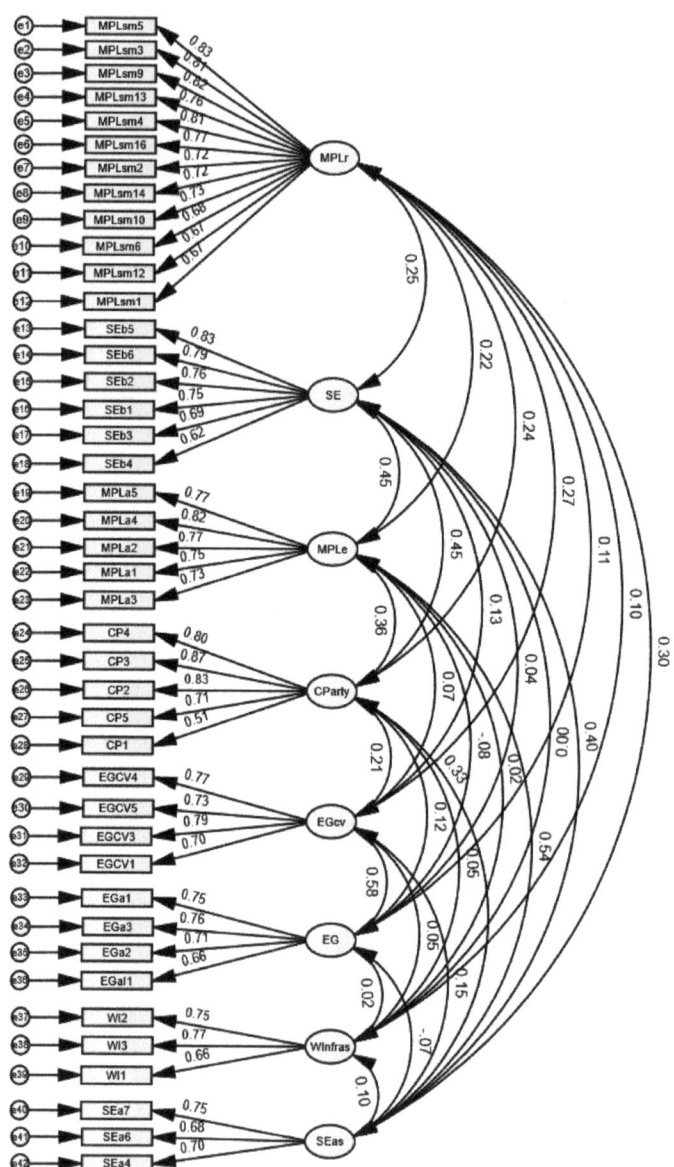

Figure 4. Measurement model of MPL reduction in Manila.

4.6.2. Measurement Model

The critical ratio (C.R.) value describes the statistics established by "dividing an estimate by its standard error". In this study, since a sample size of 426 Manila barangays is adequate for CFA, the critical ratio resembles a normal distribution. In that case, a value of 1.96 indicates two-sided significance at the "standard" 5% level. The null hypothesis is rejected since the critical ratio (CR) for a regression weight is more than (>) 1.96, indicating that the path is significant at the 0.05 level, indicating that all the estimates (for respective latent factors to the manifest variables) were statistically different from zero, as indicated in Table S3: Results of the measurement model (CFA) for the manifest and latent variables.

All of the parameter estimates were positive and within the allowed range of values of 1.00; these corresponding manifest variables were all significant at $p < 0.001$. The path coefficient for the latent factor (MPLr) to the 12 manifest variables, with standardized regression weights, was within the range of 0.653 to 0.816 (as illustrated in Figure 5 and Table S3). These results, according to [82] Hair, J., Sarstedt, M., Hopkins, L., and G. Kuppelwieser (2014), established the validity and reliability of the manifest variables. The path coefficient for the SE latent factor to the 6 manifest variables and the standardized regression weights were within the range of 0.631 to 0.848. These results, according to Hair et al., (2014) [82] established the validity and reliability of the manifest variables. The path coefficient for the MPLe latent factor to the manifest variables was significant at $p < 0.001$ and the standardized regression weights were within the range of 0.658 to 0.820. The path coefficient for CParty latent factor to the manifest variables (CP4, CP3, CP2, CP5, CP1), and the standardized regression weights were within the range of 0.514 to 0.870. The path coefficient for the EGcv latent factor to the manifest variables (EGCV4, EGCV5, EGCV3, and EGCV1), and the standardized regression weights were within the range of 0.698 to 0.792. The path coefficient for the EG latent factor to the manifest variables and the standardized regression weights were within the range of 0.658 to 0.763. The path coefficient for the WInfras latent factor to the manifest variables and the standardized regression weights were within the range of 0.655 to 0.769. The path coefficient for the SEas latent factor to the manifest variables (SEa7, SEa6, SEa4) was within the range of 0.683 to 0.750 (as illustrated in Table S3). These results, according to Hair, J., Sarstedt, M., Hopkins, L., and G. Kuppelwieser (2014) [82] established the validity and reliability of the manifest variables.

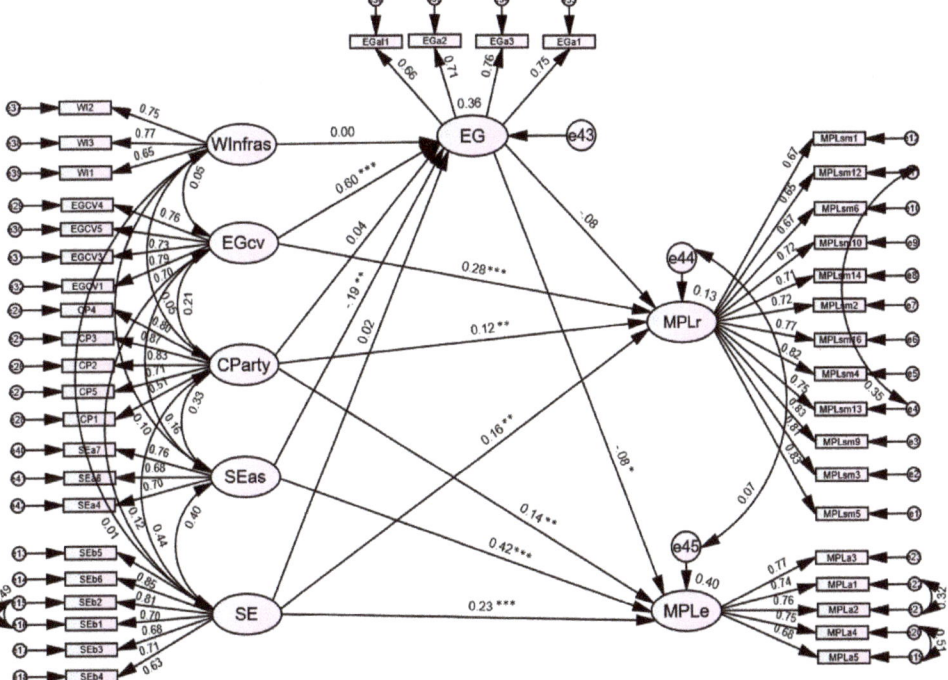

Figure 5. Structural latent model. Note: * $p < 0.050$; ** $p < 0.010$; *** $p < 0.001$.

4.6.3. Convergent and Discriminant Validity

Convergent and discriminant validity in the study was assessed using the Fornell and Larcker criterion and Heterotrait-Monotrait (HTMT) ratio. We observed the convergent and discriminant validity of the latent variables EGcv, MPLr, SE, MPLa, CParty, SEa, EG, and WInfras for our study in Manila, as evidenced by a convergent AVE of above 0.500. The discriminant is the square root of the AVE of the respective latent variables greater than the correlations, and reliability was evidenced by a composite reliability (CR) value above the threshold of 0.700. Cronbach's alpha and composite reliability were used to assess the construct reliability. Cronbach alpha was found to be higher than the required limit of 0.70 for each construct in the study [83]. Using the Fornell-Lacker criterion, diagonal elements (in bold) show the average shared-squared variance (ASV) between the latent variables and their measures (AVE). Off-diagonal elements are the correlations among latent variables. For discriminant validity, diagonal elements are larger than off-diagonal elements, as illustrated in Table 3.

Table 3. Fornell-Lacker criterion: reliability results, discriminant validity, correlation coefficient, and descriptive statistics.

	CR	AVE	EGcv	MPLr	SE	MPLe	CParty	SEas	EG	WInfras
EGcv	0.835	0.559	0.748							
MPLr	0.938	0.560	0.269 ***	0.748						
SE	0.875	0.540	0.116 *	0.239 ***	0.735					
MPLe	0.864	0.561	0.073 †	0.226 ***	0.437 ***	0.749				
CParty	0.865	0.570	0.213 ***	0.236 ***	0.443 ***	0.366 ***	0.755			
SEas	0.755	0.508	0.15 **	0.307 ***	0.402 ***	0.552 ***	0.329 ***	0.713		
EG	0.810	0.518	0.577 ***	0.110 *	0.036 †	−0.075 †	0.118 *	−0.069 †	0.719	
WInfras	0.770	0.529	0.049 †	0.100 ***	0.008 †	0.039 †	0.054 †	0.101 †	0.021 †	0.727
Cronbach's Alpha			0.904	0.957	0.879	0.885	0.913	0.780	0.900	0.791
Average Mean			3.20	3.38	3.60	3.30	3.57	3.54	2.77	2.50
Average Std. Deviation			0.67	0.65	0.53	0.64	0.57	0.55	0.78	0.95

Note: CR = composite reliability; AVE = average variance extracted; ASV = Average Shared Squared Variance. Interpretation for "Convergent Validity: CR > 0.7, CR > AVE, AVE > 0.5; for Discriminant Validity: ASV < AVE" (Fornell, C., and Larcker, 1981). Cronbach's alpha reliability coefficient > 0.7. Significance correlations: † $p > 0.100$; * $p < 0.050$; ** $p < 0.010$; *** $p < 0.001$.

While the recommendation of examining shared variance to assess discriminant validity by Fornell, C., and Larcker, (1981) [84] was once widely accepted, recent research has begun to raise questions about how sensitive this test is in capturing discriminant validity issues between constructs [85]. Following that, the heterotrait–monotrait ratio of correlations (HTMT) technique was proposed as a modern approach to determining the discriminant validity between constructs.

The HTMT method compares the correlations of indicators across constructs to the correlations of indicators within a construct, examining the ratio of between-trait correlations to the within-trait correlations of two constructs [85]. Assumptions for HTMT value should be below (<) 0.85 [86] and 0.90 [77]; discriminant validity has been established between 8 reflective constructs (as illustrated in Table 4). Hence, the current measurement model has NO validity and reliability concerns in the latent variables; internal consistency was established with good Cronbach's alpha values.

Table 4. HTMT analysis.

Construct	SEas	WInfras	EG	EGcv	CParty	MPLe	SE	MPLr
SEas								
WInfras	0.101							
EG	−0.069	0.021						
EGcv	0.150	0.049	0.578					
CParty	0.330	0.054	0.119	0.213				
MPLe	0.556	0.038	−0.084	0.070	0.356			
SE	0.396	0.008	0.035	0.114	0.438	0.434		
MPLr	0.304	0.108	0.109	0.268	0.234	0.231	0.232	

4.6.4. Assessment of Multi-Group Invariance

Group invariance was performed through configural, metric, and scalar invariance tests. We utilized the male and female groupings to test the invariance of the current measurement model. The results indicated that the configural invariance was good, as evidenced by the excellent model fit measures when estimating two groups freely, e.g., without constraints. Metric invariance was also excellent, as evidenced by a non-significant p-value of 0.567 (indicating invariant), a chi-square difference test between the unconstrained (χ^2 of 2732.194, df. of 1914) and fully constrained models (χ^2 of 2775.935, df. of 1960) where the regression weights were constrained. Scalar invariance has also an excellent result, with a p-value of 0.496 for the model measurement intercepts; hence, the current model is invariant.

4.6.5. Common Method Bias

When differences in responses are generated by the instrument rather than the actual tendencies of the respondents that the instrument aims to reveal, a common method bias (CMB) occurs, especially if the study is perpetual (e.g., opinions, perceptions). In other words, the instrument introduces a bias; hence, there are variances, which were analyzed in this research study in Manila. After conducting the validity test of the CFA, the common latent factor (CLF) was plugged into each manifest variable and the CMB was run in AMOS version 28. In the analysis of the CMB, the model fit was checked to fulfill the assumption measures. The difference between the standard regression weights of the common latent factor (CLF) in the zero-constrained model and the standard regression weights of the CLF (unconstrained model) was computed. The result difference of all regression weights is less than (<) 0.2 [87,88], indicating that there is no bias in the model in the CMB analysis.

After the measurement model confirmatory factor analysis, the outcome suggests an established composite reliability, convergent validity, discriminant validity, and no common method bias of the current measurement model. Table S3 summarizes the results of the measurement model of the 8 latent variables to the respective manifest variables. Overall, the parameter estimates or the path coefficients of the latent variables to the corresponding manifest variables were significant at $p < 0.001$.

4.6.6. Assessment of Multivariate Normality and Multicollinearity

Before proceeding with the SEM analysis, an assessment of multivariate normality and multicollinearity was conducted, using SPSS version 23. The observed values fall roughly along the straight line in the P-P plot, indicating that the observed values are similar to what we would expect from a normally distributed dataset [69]. The threshold level of the tests of skewness and kurtosis does not exceed between +2 and −2 (see Table S2). The multivariate influential value was examined using Cook's distance analysis to identify any (multivariate) existence of influential outliers. In addition, we did not observe a Cook's distance of greater than 1 [89]. Most of the case studies and Manila barangays' data points were far less than 0.05. Similarly, using multiple linear regression analysis, we examined the variance inflation factors (VIFs) for all the predictors on our dependent variables, and observed that no VIFs were greater than 2, which is far less than the threshold of 10, ensuring that we are adding unique values.

A test for autocorrelation in the residuals from a statistical model or regression study is the Durbin–Watson (DW) statistic. The Durbin–Watson statistic has a range of values from 0 to 4. A score of 2.0 implies that the sample contains no autocorrelation. A rule of thumb is that DW test statistic values in the range of 1.5 to 2.5 are relatively normal [76]. Values outside this range could, however, be a cause for concern. The outcome of the autocorrelation test of the Durbin–Watson value of the current model is 2.049, which is within the indicated range; hence, the study datasets are relatively normal.

4.7. Discussion

4.7.1. Path Model Fit

For the SEM model fit for this study in Manila, the chi-square goodness-of-fit test was not significant for the path model [78] (χ^2/df = 1.190, p-value of 0.313), suggesting that the model fits the data very well. The RMSEA was excellent at 0.021, and the pClose test was not significant (p-value of 0.730), with a GFI of 0.997, which is greater than (>) 0.95. The CFI (0.999), IFI (0.999), and TLI (0.994) were greater than (>) 0.95, while the SRMR was 0.015. All these model fit indices for the causal path model on keystone factors influencing the reduction of marine plastic litter suggest a well-fitting model. Moreover, the structural model (Figure 5) fit indices also suggest a well-fitting model, with an RMSEA of 0.035 and an SRMR of 0.046. The threshold criteria for model fit indices are in reference to [42] Hu and Bentler (1999) and [90] Bollen (1989), as depicted in Table 5. The causal path model does not apply unnatural constraints to the set of measures. Similarly, the assessment of normality was also checked in AMOS; descriptively, all the skewness and kurtosis have substantial evidence of univariate normality. The multivariate normality of the variables suggested a normalized estimate of a less than five (5) critical ratio, which is indicative of a substantial multivariate assumption [91].

Table 5. Path model fit.

Criterion of Model Fit	Absolute Fit Acceptance	Values of Model Fit	Test Result	Values of Model Fit	Test Result
		Structural Latent Model		Path Model	
p-Value for χ^2-Test	Insignificant	0.000	Significant	0.313	Insignificant
RMSEA	<0.08	0.035	Established	0.021	Established
SRMR	<0.08	0.046	Established	0.015	Established
pClose	>0.05	1.000	Established	0.730	Established
GFI	≥0.90	0.882	Acceptable	0.997	Established
AGFI	≥0.90	0.866	Acceptable	0.975	Established
CFI	≥0.90	0.955	Established	0.999	Established
IFI	≥0.90	0.955	Established	0.999	Established
TLI	≥0.90	0.951	Established	0.994	Established
χ^2/df	<3.00	1.524	Established	1.190	Established

Note: Root mean square estimation approximation (RMSEA), standardized root mean square residual (SRMR), goodness-of-fit index (GFI), adjusted goodness-of-fit index (AGFI), a comparative fit index (CFI), incremental fit index (IFI), and Tucker–Lewis index (TLI).

4.7.2. Structural Model Squared Multiple Correlations

Chinn (1998) [92] recommended R^2 values for endogenous latent variables, based on: 0.67 (substantial), 0.33 (moderate), and 0.19 (weak). Moreover, Cohen's (1988) [93] standard interpretation suggested that in terms of squared multiple correlations in SEM, "R-squared values of 0.12 or below indicate low (week effect size), between 0.13 to 0.25 values indicate medium, 0.26 or above, and above values indicate high effect size (substantial)". However, Falk, R.F., and Miller (1992) [94] recommended that R^2 values should be equal to or greater than 0.10 for the variance of a particular endogenous construct to be deemed adequate.

The corresponding R^2 of the three endogenous latent factors, EG, MPLr, and MPLe, of the *structural model* are 0.36, 0.13, and 0.40—which suggests a medium to high effect size.

Moreover, as depicted in Figure 6, the path model showed endogenous variables for EG, MPLr, and MPLe, with medium to high effect size R^2 values of 0.48, 0.19, and 0.52, respectively. Here, 52% of the variance of MPLe is explained by five exogenous variables: EG, EGcv, CParty, SEas, and SE.

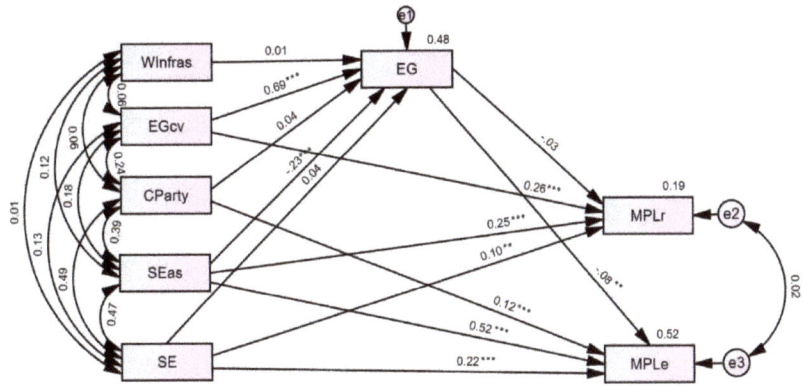

Figure 6. Path model. Note: ** $p < 0.010$; *** $p < 0.001$.

For the SEM path analysis, the study encapsulated the path coefficients, standard error, *t*-values (critical ratio), and significant *p*-values, using AMOS version 28, from standardized factor score weights in the validated CFA analysis with natural constraints, as illustrated in Table 6.

Table 6. Parameter estimates of the path model.

Causal Relationship			Standardized β	Unstandardized β	S.E.	C.R.	p-Value
WInfras	→	EG	0.010	0.009	0.031	0.287	0.774
EGcv	→	EG	0.685	0.832	0.044	18.873	0.001
CParty	→	EG	0.040	0.050	0.052	0.953	0.341
SE	→	EG	0.039	0.051	0.056	0.900	0.368
SEas	→	EG	−0.231	−0.364	0.065	−5.627	0.001
EG	→	MPLr	−0.035	−0.030	0.052	−0.577	0.564
SEas	→	MPLe	0.521	0.597	0.045	13.292	0.001
EG	→	MPLe	−0.084	−0.061	0.025	−2.456	0.014
CParty	→	MPLe	0.120	0.109	0.036	3.029	0.002
SE	→	MPLe	0.215	0.205	0.039	5.238	0.001
SE	→	MPLr	0.105	0.117	0.056	2.102	0.036
EGcv	→	MPLr	0.262	0.273	0.063	4.293	0.001
SEas	→	MPLr	0.253	0.341	0.070	4.877	0.001

4.7.3. Mediation Analysis

The study assessed the mediating role of environmental governance (EG) on the relationship between EG regarding COVID-19 waste management (EGcv) and MPL leakage (MPLe) in Manila. The results of the bootstrapped test (2000 samples) revealed that the significant indirect effect of the impact of EGcv on MPLe was negative and significant (b = −0.051, t = −2.318, *p*-value = 0.023), supporting Hypothesis 11 (H11). Furthermore, the direct effect of EGcv on EG in the presence of a mediator was also found to be significant (b = 0.832, 0.001). Hence, EG partially mediated the relationship between EGcv and MPLe. A mediation analysis summary is presented in Table 7. Similarly, regarding the mediating role of EG on the relationship between socio-economic activities (SEas—socio-economic activities) and MPL leakage, the results revealed that the significant indirect effect of the impact of SEas on MPL leakage was positive and significant (b = 0.022, t = 2, p-value = 0.018), supporting Hypothesis 12 (H12). Furthermore, the direct effect of SEas on EG in the presence of the mediator was also found to be significant (b = 0.597, 0.001). Hence, EG partially mediated the relationship between SEas and MPLe.

Table 7. Mediation results.

Relationship	Direct Effect (Unstd. β (p-Value))	Indirect Effect	Confidence Interval		p-Value	Interpretation
			Lower Bound	Upper Bound		
EGcv → EG → MPLe	0.832 (0.001)	−0.051	−0.095	−0.008	0.023	Partial mediation
SEas → EG → MPLe	0.597 (0.001)	0.022	0.004	0.047	0.018	Partial mediation

4.7.4. Multigroup Analysis

From the constrained structural weights of multigroup analysis, the global chi-squared difference test for the current model was significant at a 90% confidence level (p-value = 0.098); we observed that the model is different between males and females. The p-value means that male and female reactions are different; they do not hold the same opinions. In Table 8, we have computed the group with a Z-score as an indication of significance. The effect size of R^2 for MPLe is large, with 0.4324 (f-squared) from an R^2 difference between 0.63 (male) and 0.47 (female) [95,96]. The standardized coefficient for males is −0.24, while the coefficient for females is −0.21. Hence, a positive relationship between socio-economic activities (SEas) and EG is stronger for females, supporting Hypothesis 15 (H15).

Table 8. Multigroup differences (male and female).

Causal Path Relationship			Male		Female		z-Score
			Estimate	p-Value	Estimate	p-Value	
WInfras	→	EG	−0.022	0.676	0.024	0.540	0.701
EGcv	→	EG	0.767	0.000	0.878	0.000	1.250
CParty	→	EG	0.019	0.823	0.069	0.295	0.466
SE	→	EG	−0.036	0.691	0.085	0.233	1.049
SEas	→	EG	−0.208	0.053	−0.432	0.000	−1.658 *
EG	→	MPLr	0.084	0.350	−0.080	0.208	−1.488
SEas	→	MPLe	0.602	0.000	0.595	0.000	−0.075
EG	→	MPLe	−0.113	0.003	−0.037	0.245	1.523
CParty	→	MPLe	0.121	0.023	0.105	0.026	−0.225
SE	→	MPLe	0.194	0.000	0.207	0.000	0.164
SE	→	MPLr	0.088	0.351	0.134	0.055	0.392
EGcv	→	MPLr	0.268	0.007	0.261	0.001	−0.050
SEas	→	MPLr	0.436	0.000	0.281	0.001	−1.060

Note: * p-value < 0.10.

4.7.5. Discussion of the Path Model Results

Our findings revealed that waste infrastructure (WInfras) is not a predictor of environment governance (EG), with a path coefficient of 0.010. A weak positive correlation between the two variables was non-significant at a p-value of 0.774. as illustrated in Tables 6 and 9. This implies that there is no relationship with waste infrastructure (WInfras) on EG; therefore, Hypothesis 1 (H1) is not supported. The result of this latent factor is not supported, due perhaps to the limited specific instrument or manifest variables under study. However, from the interview and FGD, Manila LGU requires robust evidence-based baseline data and information appropriate for policy; once the baseline data are established, they can be used to select the appropriate facilities/infrastructure and technologies in addressing MPL reduction, with the private sector's involvement in SWM. It was argued that studies on improving waste management infrastructure would necessitate significant investments (and time), particularly in the least developed and developing economies and cities and that these countries' primary focus should be on improving solid waste collection and management [7]. From the interviews and FGD, the city of Manila currently does not have a material recovery facility (MRF), particularly in the barangays; this is because of a lack of space to house a city MRF. The city instead mandated that every public school should practice a waste reduction scheme and resource recovery and should designate an area where used/soiled school and office papers, PET bottles, and old newspapers can be stored and eventually sold to a nearby junk shop for recycling.

Table 9. Hypothesis analytics.

Hypothesis Path Relationship	Coefficient	t-Value	Interpretation
H1: WInfras → EG	0.010	0.287	Not Supported
H2: EGcv → EG	0.685	18.873 ***	Supported
H3: EGcv → MPLr	0.262	4.293 ***	Supported
H4: CParty → EG	0.040	0.953	Not Supported
H5: CParty → MPLe	0.120	3.029 **	Supported
H6: SEas → EG	−0.231	−5.627 ***	Supported
H7: SEas → MPLr	0.253	4.877 ***	Supported
H8: SEas → MPLe	0.521	13.292 ***	Supported
H9: SE → MPLr	0.105	2.102 *	Supported
H10: SE → MPLe	0.215	5.238 ***	Supported
H11: EGcv → EG → MPLe (Mediation)	−0.051	−2.318 *	Supported
H12: SEas → EG → MPLe (Mediation)	0.022	2.000 *	Supported
H13: EG → MPLr	−0.035	−0.577	Not Supported
H14: EG → MPLe	−0.084	−2.456 **	Supported
H15: SEas → EG (Multigroup)	−0.208	−5.374 ***	Supported

Note: *** p-value < 0.001; ** p-value < 0.01; * p-value < 0.05.

Environmental governance regarding the "management of COVID-19-related healthcare waste" (EGcv) has a positive effect on existing environmental governance (EG). The result supports Hypothesis 2 (H2), which predicted a relationship, and implies that EGcv is a predictor for EG, with a strong and positive path coefficient of 0.685 (t = 18.873 ***). Environmental governance regarding the "management of COVID-19-related healthcare waste" (EGcv) has a positive effect on marine plastic litter solution measures (MPLr), with a positive path coefficient of 0.262 (t = 4.293 ***). This result supports Hypothesis 3 (H3), which predicted a positive relationship between the two variables. This is evidenced by the COVID-19 waste management interim guidelines set by the central government to the cities/LGUs in the Philippines for the proper management of related COVID-19 waste, e.g., proper waste segregation at the source and appropriate disposal. Most of the health care waste was managed by waste service providers. However, most of the barangays in Manila were not aware or oriented with the existing COVID-19 waste guidelines.

Moreover, community participation (CParty) is not a predictor of environment governance (EG), with an insignificant path coefficient of 0.040. This implies that there is no relationship between community participation (CParty) and EG, which does not support Hypothesis 4 (H4). However, community participation (CParty) is positively related to marine plastic litter leakage (MPLe), with a positive path coefficient of 0.120 (t = 3.029 **), supporting Hypothesis 5 (H5). This is evident in the current participation of the city SWM board, the city, the barangay, private entities and institutions, citizens, NGOs, and recycling companies in implementing effective SWM and marine plastic litter prevention. The city, in collaboration with the Manila Barangay Bureau, the Division of City Schools, the Manila Health Department, and the Bureau of Permits and Licensing Office, has launched a massive information, education, and communication (IEC) campaign targeting all sectors and generations to instill the basic requirements of waste segregation at source and segregated storage, pending collection. Moreover, the barangays are required to guarantee that their constituents follow the Barangay Solid Waste Management Committee's ordinance to conduct waste segregation at the source and discourage the illegal disposal of household waste. However, there is a need for participation upstream by plastic industries stakeholders for MPL reduction. The authors believe that demand drives marine plastic waste in the Philippines. It was observed that researchers must follow the terms of reference of the development partners/donors in the MPL project. In addition, there is a weak relationship between science and policy in the Philippines, with most of the research conducted by local scholars ending up on the shelf rather than being translated into policy. The relationship between the government and academia should be enhanced.

The socio-economic activities factor (SEas) is a predictor of existing environmental governance (EG), with a path coefficient of -0.231 ($t = -5.627$ ***), and was found to be significant, which means that there is a correlation between the SEas and EG variables. Hence, Hypothesis 6 (H6) is supported. In the same way, socio-economic activities (SEas) have a positive association with marine plastic litter solution measures (MPLr), with a path coefficient of 0.253 ($t = 4.877$ ***), supporting Hypothesis 7 (H7). Socio-economic activities (SEas) have a positive association with marine plastic litter leakage (MPLe), with a path coefficient of 0.521 (13.292 ***), supporting Hypothesis 8 (H8). It should be underlined that the Manila government and the Philippine LGUs consider several socio-economic drivers when addressing the current practical challenges of marine plastic litter abatement. These drivers include the domestic and international market and economic forces, legislation, the design of products and services, urbanization and consumerism patterns, regional cooperation, and human behavior and convenience factors that affect sustainable consumption and production. Nevertheless, this is evident in the current SWM implementation in Manila through the Department of Public Services (DPS), which continues to send open letters to businesses to encourage them to practice source separation and to support the city's environmental awareness program by maintaining the cleanliness of their surroundings and ensuring that no waste produced during their operations is disposed of in canals or estuaries that lead to the Pasig River. There are 17 major river systems that drain into Manila Bay, and the rivers are home to informal settler families. The DENR-led Manila Bay Cleanup Program, in cooperation with the Manila LGU, undertakes the following actions in compliance with the Writ of Continuing Mandamus: clean-up for water quality improvement, rehabilitation and resettlement, and education and sustainability. The DENR conducted 2025 clean-up drives with 25,595 volunteers in the fourth quarter of 2020, collecting and disposing of 1406 tons of waste, on top of activities by the PNP-MG, MMDA, and local government. The program also addresses informal settlements (sourced during an interview with Department of Environment and Natural Resources (DENR), 2021). With limited resources, frequent clean-up activities may not be sustainable since they entail substantial financial and human resources. However, Manila could showcase good practices of LGU environmental governance, to replicate the model in other neighboring cities to Metro Manila, e.g., the DPS Beach Warriors and estuary rangers cleaning up the city of Manila; initiatives such as LinISKOmaynila and Aling Tindira (the vendor waste-to-cash program), in partnership with the Coca-Cola company; most of the estuaries or rivers have screen traps; however, these artificial barriers may not be sustainable since they will be prone to damage during flash-flooding.

Likewise, Manila public litter behavior in terms of the MPL problem (SE) has a positive effect on marine plastic litter solution measures (MPLr), with a path coefficient of 0.105 ($t = 2.102$ *), supporting Hypothesis 9 (H9). Manila's public behavior regarding the MPL problem (SE) has a positive association with marine plastic litter leakage (MPLe), with a path coefficient of 0.215 ($t = 5.238$ ***), supporting Hypothesis 10 (H10). SE is measured by variable indicators, such as a lack of enforcement of waste disposal directives, a lack of funding for waste collection, a lack of waste collection and separation, public behavior in terms of littering, a lack of adequate waste management infrastructure, and the intense consumption of single-use plastics. This is evident in the instrument for environmental governance (software component), which is not associated with the socio-economic activities factor. This also suggests the enforcement of SWM plans and cleanliness.

However, the mediated effects in Hypotheses H11 and H12 are supported, which implies that environmental governance is an intervening variable that partially relates to the mediated variables. The bootstrapped test for mediation for Hypothesis 11 (H11) revealed a significant indirect effect of the impact of EGcv on MPLe, which was negative and significant with a path coefficient of -0.051 ($t = -2.318$ *), suggesting that EG has partially mediated the relationship between EGcv and MPLe. Likewise, the mediating role of EG on the relationship between SEas and MPLe revealed a significant indirect effect of the impact of SEas on MPLe, which was positive and significant, with a path coefficient of 0.022 ($t = 2.0$ *), supporting Hypothesis 12 (H12)—implying that EG partially mediated the relationship between SEas and MPLe.

For environmental governance (e.g., SWM and MPL policy, strategy, and guidelines) the factor of EG is not a predictor of marine plastic litter solution measures, with a standard regression coefficient of -0.035 and a negative correlation between the two variables (latent and manifest variables) that was found to be insignificant. This suggests that enabling environmental governance (EG) may not be potentially correlated with MPL solution measures. This is evident in the instrument used in the study, as manifested by the measures of variable indicators for environmental governance (software), which described the need for clear guidelines and strategy for MPL/SWM; effective mechanisms in place for the waste facility; the openness, transparency, and accountability of bid processes in SWM; and public involvement at appropriate stages of the SWM decision-making. In contrast, the MPL solution measures entail mostly physical aspects (hardware), such as enhancing waste separation at all households and establishments, establishing material recycling facilities, awareness-raising campaigns, producing packaging made from alternative materials, and enhancing waste collection coverage in barangays, amongst other measures. The result does not support Hypothesis 13 (H13). Our findings, however, contradict the works of Scheinberg, A., Wilson, D.C., and Rodic, (2010) [97] showing overlapping components of integrated solid waste management and the physical and governance components; the reason for the difference might be due to the methodological techniques.

Environmental governance, in terms of strategies, guidelines, and implementation procedures (EG), positively impacts marine plastic litter leakage (MPLe), with a path coefficient of -0.084 ($t = -2.456$ **), supporting Hypothesis 14 (H14). This is evident in the existing scenario, where implementation progresses the 10-year Manila SWM plan and the developed roadmap for preventing marine plastic litter in Manila—in alignment with the approved national plan of action on marine litter prevention in the Philippines. Moreover, the existence of environmental governance, political will, and transparency are among the driving forces that spur effective waste management in Manila, which can be observed in the current administration. As mandated by R.A. 9003 (the Philippine ESWM Act), the DPS conducts IEC in every barangay to encourage source segregation and waste reduction. In addition, Manila is progressing in terms of implementing and enforcing its SWM strategies and plans through its existing segregation strategies. In addition, information, education, and communication (IEC), which are being promoted by the Manila DPS—e.g., source separation at the barangay level—is one of the environmental governance practices that the city is implementing, following their SWM strategy and plans. However, Manila is yet to institutionalize an SWM department to tackle the SWM in the city. For the city's SWM spending, a total of PHP 602,588,348.00 (USD 12,297,721.39) was budgeted in 2015 for the collection and disposal of solid waste, including street-sweeping services and IEC activities. However, if the barangays encouraged recycling, the city's SWM spending could be reduced and may make savings.

For the multigroup effect, the positive relationship between socio-economic activities (SEas) and MPL leakage (MPLe) is stronger for females, as evidenced by the path coefficients of −0.24 (male) and −0.21 (females, $t = -5.374$ ***); a global chi-squared difference test for the current model was significant at a 90% confidence level. Hence, we observed that the model is different according to gender (the *p*-value of males and females are different), which supports Hypothesis 15 (H15). These findings and results provided significant perspectives for Manila for enabling environmental governance for marine plastic reduction (illustrated in Table 9, Hypothesis analytics).

4.7.6. Bollen–Stine Bootstrap Test

Using bootstraps in AMOS version 28, the researcher performed a Bollen–Stine bootstrap test with 5000 bootstrap samples [98]. The result of the Bollen–Stine bootstrap test has a non-significant *p*-value of 0.320, suggesting that the current model is correct.

5. Conclusions

The keystone factors regarding the practical challenges, including socio-economic activities, environmental governance, community participation, waste infrastructure, and solution measures for marine plastic litter (MPL) reduction in Manila City, are cross-cutting and are related to the SDGs. Consequently, the data fits well with the measurement model, which suggests an established construct validity and reliability. Moreover, the structural model has established good test result fit measures, including the *p*-value = 0.313, RMSEA = 0.021, SRMR = 0.015, GFI = 0.997, CFI = 0.999, TLI = 0.994, IFI = 0.999, and the chi-square/degrees of freedom ratio = 1.190. We found that the environmental governance of SWM policies and guidelines (EG) and COVID-19 waste management (EGcv), community participation, socio-economic activities, and the public litter behavior factor have positively influenced MPL leakage and MPL solution measures. In addition, environmental governance (SWM policies and guidelines) mediating EGcv and SEas positively impacts MPL leakage. However, there is no relationship between waste infrastructure and environmental governance; Manila LGU has limited resources to effectively implement the existing national strategies and actions including the installation of waste infrastructure, e.g., appropriate artificial screen traps and MRF. Manila has a burden of pollution, and the challenge of current waste management is great, in particular, waste separation at source and the lack of a disposal facility, as the city relies on landfills in a neighboring city. Manila City also has a challenge related to healthcare waste collection, treatment, transportation, and disposal, as seen during the first year of the COVID-19 pandemic. Direct anthropogenic activities, such as population growth, urbanization, tourism, inward migration, intense plastic production [32,35,36,64], and slum residents along the river/canals and coastal areas are drivers that exacerbate marine plastic littering within Manila and in the Philippine context.

There is a great need to empower the barangays so that they can play a part in preventing and reducing marine plastic litter from leaking into the ocean. In this context, opportunities for showcasing dynamic innovation in Manila should be considered. In the Philippines, Manila and other LGUs should disseminate a wide range of practices that move away from the traditional linear (take-make-use-dispose) way of thinking [22], allowing barangays and LGUs to be more responsible in terms of achieving long-term waste management and circularity. However, due to COVID-19 pandemic restrictions, only a modest amount of baseline data was collected. Future research could include the plastics industry and informal recyclers, as well as the use of longitudinal data from Philippine cities/municipalities, combining an SEM and empirical DPSIR approach. The study would improve the efficacy of using SEM systematically and inclusively.

Supplementary Materials: The following are available online at https://www.mdpi.com/article/10.3390/su14106128/s1, Table S1. Demographic characteristics of respondents. Table S2. Factors in MPL Reduction with mean, SD, Skewness, Kurtosis, factor loadings and Cronbach's Alpha. Table S3. Results of Measurement Model (CFA) for the manifest and latent variables.

Author Contributions: Conceptualization, G.B.; data curation, G.B.; formal analysis, G.B.; investigation, G.B.; methodology, G.B.; supervision, A.N.; validation, A.N.; writing—original draft, G.B.; writing—review and editing, G.B. and A.N. All authors have read and agreed to the published version of the manuscript.

Funding: This research received no external funding.

Institutional Review Board Statement: National Institute of Development Administration (protocol code ECNIDA 2021/0106 and 9 September 2021).

Informed Consent Statement: If you decide to participate in the research project, you will be asked to complete the questionnaire that will ask you some questions about the status of municipal solid waste and marine plastic litter even during the current COVID-19 pandemic - to identify gaps in waste management, local realities and draw out sustainable waste management solutions. The questionnaire contains more than 50 questions in 4 likert scale, divided into 10 parts/sections. We estimate that the questionnaire will take you about 20 min in one seating to complete. You are under no obligation to complete questions that you would prefer not to answer.

Data Availability Statement: Not applicable.

Acknowledgments: This research was supported by the National Institute of Development Administration, Thailand. The authors would like to acknowledge and thank the endorsement of the Manila Department of Public Services in the conduct of the survey, as well as for the interviews and focused group discussions.

Conflicts of Interest: The authors declare no conflict of interest. The authors declare that they have no known competing financial interests or personal relationships that could have appeared to influence the work reported in this paper.

References

1. The Pew Charitable Trusts. *Breaking the Plastic Wave*; The Pew Charitable Trusts: Philadelphia, PA, USA, 2020; Volume 56. Available online: https://www.pewtrusts.org/-/media/assets/2020/07/breakingtheplasticwave_summary.pdf (accessed on 25 March 2022).
2. Law, K.L.; Starr, N.; Siegler, T.R.; Jambeck, J.R.; Mallos, N.J.; Leonard, G.H. The United States' contribution of plastic waste to land and ocean. *Sci. Adv.* **2020**, *6*, eabd0288. [CrossRef] [PubMed]
3. Lebreton, L.C.M.; Van Der Zwet, J.; Damsteeg, J.W.; Slat, B.; Andrady, A.; Reisser, J. River plastic emissions to the world's oceans. *Nat. Commun.* **2017**, *8*, 15611. [CrossRef] [PubMed]
4. Jambeck, J.R.; Geyer, R.; Wilcox, C.; Siegler, T.R.; Perryman, M.; Andrady, A.; Narayan, R.; Law, K.L. Plastic waste inputs from land into the ocean. *Science* **2015**, *347*, 768–771. [CrossRef] [PubMed]
5. Willis, K.; Maureaud, C.; Wilcox, C.; Hardesty, B.D. How successful are waste abatement campaigns and government policies at reducing plastic waste into the marine environment? *Mar. Policy* **2018**, *96*, 243–249. [CrossRef]
6. Van Assche, K.; Beunen, R.; Duineveld, M. *Evolutionary Governance Theory: An Introduction*; Springer: Berlin/Heidelberg, Germany, 2014. [CrossRef]
7. Löhr, A.; Savelli, H.; Beunen, R.; Kalz, M.; Ragas, A.; Van Belleghem, F. Solutions for global marine litter pollution. *Curr. Opin. Environ. Sustain.* **2017**, *28*, 90–99. [CrossRef]
8. UNEP Factsheet Series. In *Environmental Governance*; UNEP: Vienna, Austria. Available online: http://www.unep.org/environmentalgovernance/ (accessed on 2 July 2020).
9. UNEP. *Resolution 3/4-United Nations Environment Assembly of the United Nations Environment Programme*; United Nations Environment Programme: Nairobi, Kenya, 2022; pp. 1–6. Available online: https://papersmart.unon.org/resolution/uploads/k1900699.pdf (accessed on 25 March 2022).
10. Lyons, B.P.; Cowie, W.J.; Maes, T.; Le Quesne, W.J.F. Marine plastic litter in the ROPME Sea Area: Current knowledge and recommendations. *Ecotoxicol. Environ. Saf.* **2020**, *187*, 109839. [CrossRef] [PubMed]
11. Yang, Y.; Chen, L.; Xue, L. Chinese Journal of Population, Resources and Environment Looking for a Chinese solution to global problems: The situation and countermeasures of marine plastic waste and microplastics pollution governance system in China. *Chin. J. Popul. Resour. Environ.* **2022**, *19*, 352–357. [CrossRef]
12. Kulkarni, B.N.; Anantharama, V. Repercussions of COVID-19 pandemic on municipal solid waste management: Challenges and opportunities. *Sci. Total Environ.* **2020**, *743*, 140693. [CrossRef]
13. Klemeš, J.J.; Van Fan, Y.; Tan, R.R.; Jiang, P. Minimising the present and future plastic waste, energy and environmental footprints related to COVID-19. *Renew. Sustain. Energy Rev.* **2020**, *127*, 109883. [CrossRef]
14. Bugge, M.M.; Fevolden, A.M.; Klitkou, A. Governance for system optimization and system change: The case of urban waste. *Res. Policy* **2019**, *48*, 1076–1090. [CrossRef]

15. Soltani, A.; Sadiq, R.; Hewage, K. The impacts of decision uncertainty on municipal solid waste management. *J. Environ. Manag.* **2017**, *197*, 305–315. [CrossRef] [PubMed]
16. Ocean Conservancy. *The Next Wave: Investment Strategies for Plastic Free Seas*; Ocean Conservancy: Washington, DC, USA, 2017.
17. Garnett, K.; Cooper, T.; Longhurst, P.; Jude, S.; Tyrrel, S. A conceptual framework for negotiating public involvement in municipal waste management decision-making in the UK. *Waste Manag.* **2017**, *66*, 210–221. [CrossRef] [PubMed]
18. Benson, N.U.; Fred-Ahmadu, O.H.; Bassey, D.E.; Atayero, A.A. COVID-19 pandemic and emerging plastic-based personal protective equipment waste pollution and management in Africa. *J. Environ. Chem. Eng.* **2021**, *9*, 105222. [CrossRef] [PubMed]
19. Shiong, K.; Yiing, L.; Ren, H.; Yi, H.; Wayne, K. Plastic waste associated with the COVID-19 pandemic: Crisis or opportunity? *J. Hazard. Mater.* **2021**, *417*, 126108. [CrossRef]
20. Benson, N.U.; Bassey, D.E.; Palanisami, T. COVID pollution: Impact of COVID-19 pandemic on global plastic waste footprint. *Heliyon* **2021**, *7*, e06343. [CrossRef]
21. GESAMP. Guidelines for the monitoring and assessment of plastic litter in the ocean. In *Reports and Studies*; Kershaw, P.J., Turra, A., Galgani, F., Eds.; IMO/FAO/UNESCO-IOC/UNIDO/WMO/IAEA/UN/UNEP/UNDP/ISA Joint Group of Experts on the Scientific Aspects of Marine Environmental Prote; GESAMP: London, UK, 2019; Volume 99. Available online: http://www.gesamp.org/publications/guidelines-for-the-monitoring-and-assessment-of-plastic-litter-in-the-ocean (accessed on 25 March 2022).
22. GIZ. *Marine Litter Prevention: Reducing Plastic Waste Leakage into Waterways and Oceans through Circular Economy and Sustainable Waste Management*; GIZ: Berlin, Germany, 2018.
23. PREVENT Waste Alliance. *EPR Toolbox: Know-How to Enable Extended Producer Responsibility for Packaging*; PREVENT Waste Alliance: Bonn, Germany, 2020; pp. 1–205. Available online: https://prevent-waste.net/en/epr-toolbox/ (accessed on 25 March 2022).
24. Wilson, D.C.; Rodic, L.; Cowing, M.J.; Velis, C.A.; Whiteman, A.D.; Scheinberg, A.; Vilches, R.; Masterson, D.; Stretz, J.; Oelz, B. "Wasteaware" benchmark indicators for integrated sustainable waste management in cities. *Waste Manag.* **2015**, *35*, 329–342. [CrossRef]
25. Oke, A.; Osobajo, O.; Obi, L.; Omotayo, T. Rethinking and optimising post-consumer packaging waste: A sentiment analysis of consumers' perceptions towards the introduction of a deposit refund scheme in Scotland. *Waste Manag.* **2020**, *118*, 463–470. [CrossRef]
26. Ryberg, M.W.; Laurent, A.; Hauschild, M. *Mapping of Global Plastics Value Chain and Plastics Losses to the Environment (with a Particular Focus on Marine Environment)*; UN Environ: Nairobi, Kenya, 2018; pp. 1–99.
27. Soria, K.Y.; Palacios, M.R.; Morales Gomez, C.A. Governance and policy limitations for sustainable urban land planning. The case of Mexico. *J. Environ. Manag.* **2020**, *259*, 109575. [CrossRef]
28. Carman, V.G.; Machain, N.; Campagna, C. Legal and institutional tools to mitigate plastic pollution affecting marine species: Argentina as a case study. *Mar. Pollut. Bull.* **2015**, *92*, 125–133. [CrossRef]
29. Alpizar, F.; Carlsson, F.; Lanza, G.; Carney, B.; Daniels, R.C.; Jaime, M.; Ho, T.; Nie, Z.; Salazar, C.; Tibesigwa, B.; et al. A framework for selecting and designing policies to reduce marine plastic pollution in developing countries. *Environ. Sci. Policy* **2020**, *109*, 25–35. [CrossRef]
30. Adam, I.; Walker, T.R.; Clayton, C.A.; Carlos Bezerra, J. Attitudinal and behavioural segments on single-use plastics in Ghana: Implications for reducing marine plastic pollution. *Environ. Chall.* **2021**, *4*, 100185. [CrossRef]
31. Gall, S.C.; Thompson, R.C. The impact of debris on marine life. *Mar. Pollut. Bull.* **2015**, *92*, 170–179. [CrossRef] [PubMed]
32. Binetti, U.; Silburn, B.; Russell, J.; van Hoytema, N.; Meakins, B.; Kohler, P.; Desender, M.; Preston-Whyte, F.; Fa'abasu, E.; Maniel, M.; et al. First marine litter survey on beaches in Solomon Islands and Vanuatu, South Pacific: Using OSPAR protocol to inform the development of national action plans to tackle land-based solid waste pollution. *Mar. Pollut. Bull.* **2020**, *161*, 111827. [CrossRef] [PubMed]
33. Plastics Europe. Plastics–the facts 2018. In *An Analysis of European Plastics Production, Demand and Waste Data*; Plastics Europe: Brussels, Belgium, 2018.
34. Geyer, R.; Jambeck, J.R.; Law, K.L. Production, use, and fate of all plastics ever made. *Sci. Adv.* **2017**, *3*, 25–29. [CrossRef]
35. Davison, S.M.C.; White, M.P.; Pahl, S.; Taylor, T.; Fielding, K.; Roberts, B.R.; Economou, T.; Mcmeel, O.; Kellett, P.; Fleming, L.E. Public concern about, and desire for research into, the human health effects of marine plastic pollution: Results from a 15-country survey across Europe and Australia. *Glob. Environ. Chang.* **2021**, *69*, 102309. [CrossRef]
36. Nelms, S.E.; Eyles, L.; Godley, B.J.; Richardson, P.B.; Selley, H.; Solandt, J.L.; Witt, M.J. Investigating the distribution and regional occurrence of anthropogenic litter in English marine protected areas using 25 years of citizen-science beach clean data. *Environ. Pollut.* **2020**, *263*, 114365. [CrossRef]
37. Oliver, C. Strategic Responses to Institutional Processes. *Acad. Manag. Rev.* **1991**, *16*, 147–179. [CrossRef]
38. DiMaggio, P.J.; Powell, W.W. The Iron Cage Revisited: Institutional Isomorphism & Collective Rationality in Organizational Field. *Am. Sociol. Rev.* **1983**, *48*, 147–160.
39. Oti-Sarpong, K.; Shojaei, R.S.; Dakhli, Z.; Burgess, G. How countries achieve greater use of offsite manufacturing to build new housing: Identifying typologies through institutional theory. *Sustain. Cities Soc.* **2022**, *76*, 103403. [CrossRef]
40. Bolognesi, T.; Nahrath, S. Environmental Governance Dynamics: Some Micro Foundations of Macro Failures. *Ecol. Econ.* **2020**, *170*, 106555. [CrossRef]

41. Whiteman, A.; Smith, P.; Wilson, D.C. Waste management: An indicator of urban governance. In Proceedings of the UN-Habitat Global Conference on Urban Development, New York, NY, USA, 4 June 2001.
42. Glasbergen, P. The Question of Environmental Governance. In *Co-Operative Environmental Governance*; public-private agreements as a policy strategy; Glasbergen, P., Ed.; Kluwer Academic Publishers: Dordrecht, The Netherlands, 1998.
43. Breukelman, H.; Krikke, H.; Löhr, A. Failing services on urban waste management in developing countries: A review on symptoms, diagnoses, and interventions. *Sustainability* **2019**, *11*, 6977. [CrossRef]
44. Cai, K.; Xie, Y.; Song, Q.; Sheng, N.; Wen, Z. Identifying the status and differences between urban and rural residents' behaviors and attitudes toward express packaging waste management in Guangdong Province, China. *Sci. Total Environ.* **2021**, *797*, 148996. [CrossRef] [PubMed]
45. Wu, H.H. A study on transnational regulatory governance for marine plastic debris: Trends, challenges, and prospect. *Mar. Policy* **2020**, *136*, 103988. [CrossRef]
46. Cavalletti, B.; Corsi, M. By diversion rate alone: The inconsistency and inequity of waste management evaluation in a single-indicator system. *Reg. Sci.* **2018**, *98*, 307–329. [CrossRef]
47. Citroni, G.; Lippi, A.; Profeti, S. Local public services in Italy: Still fragmentation. In *Public and Social Services in Europe*; Palgrave Macmillan: Houndmills, UK, 2019; pp. 103–117.
48. Morseletto, P. A new framework for policy evaluation: Targets, marine litter, Italy and the Marine Strategy Framework Directive. *Mar. Policy* **2020**, *117*, 103956. [CrossRef]
49. Rangel-Buitrago, N.; Williams, A.; Costa, M.F.; de Jonge, V. Curbing the inexorable rising in marine litter: An overview. *Ocean Coast. Manag.* **2020**, *188*, 105133. [CrossRef]
50. Loukil, F.; Rouached, L. Resources, Conservation and Recycling Modeling packaging waste policy instruments and recycling in the MENA region. *Resour. Conserv. Recycl.* **2012**, *69*, 141–152. [CrossRef]
51. UNEP. *MARINE PLASTIC DEBRIS Global Lessons and Research to Inspire Action*; UNEP: Vienna, Austria, 2016; pp. 1–192. [CrossRef]
52. Magni, S.; Torre CDella Garrone, G.; Amato, A.D.; Parenti, C.C.; Binelli, A. First evidence of protein modulation by polystyrene microplastics in a freshwater biological model. *Environ. Pollut.* **2019**, *250*, 407–415. [CrossRef]
53. Di, M.; Wang, J. Science of the Total Environment Microplastics in surface waters and sediments of the Three Gorges. *Sci. Total Environ.* **2018**, *616–617*, 1620–1627. [CrossRef]
54. Mani, T.; Hauk, A.; Walter, U.; Burkhardt-Holm, P. Microplastics profile along the Rhine River. *Sci. Rep.* **2015**, *5*, 17988. [CrossRef]
55. Gasperi, J.; Dris, R.; Bonin, T.; Rocher, V.; Tassin, B. Assessment of floating plastic debris in surface water along the Seine River. *Environ. Pollut.* **2014**, *195*, 163–166. [CrossRef]
56. Morritt, D.; Stefanoudis, P.V.; Pearce, D.; Crimmen, O.A.; Clark, P.F. Plastic in the Thames: A river runs through it. *Mar. Pollut. Bull.* **2014**, *78*, 196–200. [CrossRef] [PubMed]
57. Bentler, P.M.; Chou, C.P. Practical issues in structural modeling. *Sociol. Methods Res.* **1987**, *16*, 78–117. [CrossRef]
58. Manila Department of Public Services (DPS). *Socio-Economic Demography of Manila City*; Manila DPS: Manila, Philippines, 2021.
59. Manila DPS. *Manila City 10-Year SWM Plan*; Manila DPS: Manila, Philippines, 2020.
60. Abate, T.G.; Börger, T.; Aanesen, M.; Falk-Andersson, J.; Wyles, K.J.; Beaumont, N. Valuation of marine plastic pollution in the European Arctic: Applying an integrated choice and latent variable model to contingent valuation. *Ecol. Econ.* **2020**, *169*, 106521. [CrossRef]
61. Adam, I.; Walker, T.R.; Bezerra, J.C.; Clayton, A. Policies to reduce single-use plastic marine pollution in West Africa. *Mar. Policy* **2020**, *116*, 103928. [CrossRef]
62. Gari, S.R.; Ortiz Guerrero, C.E.; A-Uribe, B.; Icely, J.D.; Newton, A. A DPSIR-analysis of water uses and related water quality issues in the Colombian Alto and Medio Dagua Community Council. *Water Sci.* **2018**, *32*, 318–337. [CrossRef]
63. Krelling, A.P.; Williams, A.T.; Turra, A. Differences in perception and reaction of tourist groups to beach marine debris that can influence a loss of tourism revenue in coastal areas. *Mar. Policy* **2017**, *85*, 87–99. [CrossRef]
64. Rochman, C.M.; Cook, A.M.; Koelmans, A.A. Plastic debris and policy: Using current scientific understanding to invoke positive change. *Environ. Toxicol. Chem.* **2016**, *35*, 1617–1626. [CrossRef]
65. Plummer, R.; Armitage, D.R.; de Loë, R.C. Adaptive comanagement and its relationship to environmental Governance. *Ecol Soc.* **2013**, *18*, 21. [CrossRef]
66. 66. National Solid Waste Management Commission, DENR Philippines. NSWMC Resolution No. 1363, Series of 2020. *The Interim Guidelines on the Management of COVID-19 Related Healthcare Waste*. Available online: https://nswmc.emb.gov.ph/wp-content/uploads/2021/06/2020-NSWMC-Reso-1364-series-of-2020.pdf (accessed on 20 July 2021).
67. MacCallum, R.C.; Browne, M.W.; Sugawara, H.M. Power analysis and determination of sample size for covariance structure modeling. *Psychol. Methods* **1996**, *1*, 130–149. [CrossRef]
68. Choi, H.S.C.; Sirakaya, E. Measuring Residents' Attitude toward Sustainable Tourism: Development of Sustainable Tourism Attitude Scale. *J. Tour. Res.* **2005**, *43*, 380–394. [CrossRef]
69. Field, A. *Discovering statistics Using SPSS for windows*; SAGE Publications: Thousand Oaks, CA, USA, 2000.
70. Field, A. *Discovering statistics Using SPSS*; Sage Publications Limited: London, UK, 2009.
71. Ferguson, E.; Cox, T. Exploratory factor analysis: A users' guide. *Int. J. Sel. Assess.* **1993**, *1*, 84–94. [CrossRef]
72. Reio, T.G., Jr.; Shuck, B. Exploratory factor analysis: Implications for theory, research, and practice. *Adv. Dev. Hum. Resour.* **2015**, *17*, 12–25. [CrossRef]

73. Hair, J.F. *Multivariate Data Analysis*; Prentice Hall: Upper Saddle River, NJ, USA, 2005.
74. Hair, J.F.; Black, W.C., Jr.; Babin, B.J.; Anderson, R.E. *Multivariate Data Analysis*, 7th ed.; Prentice-Hall: Upper Saddle River, NJ, USA, 2010.
75. Fabrigar, L.R.; Wegener, D.T.; MacCallum, R.C.; Strahan, E.J. Evaluating the use of exploratory factor analysis in psychological research. *Psychol. Methods* **1999**, *4*, 272. [CrossRef]
76. Tabachnick, B.G.; Fidell, L.S.; Ullman, J.B. *Using Multivariate Statistics*, 6th ed.; Pearson: Boston, MA, USA, 2014.
77. Hu, L.T.; Bentler, P.M. Cutoff criteria for fit indexes in covariance structure analysis: Conventional criteria versus new alternatives. *Struct. Equ. Model. A Multidiscip. J.* **1999**, *6*, 1–55. [CrossRef]
78. Bagozzi, R.P.; Yi, Y. Specification, evaluation and interpretation of structural equation models. *J. Acad. Mark. Sci.* **2012**, *40*, 8–34. [CrossRef]
79. Baumgartner, H.; Homburg, C. Applications of Structural Equation Modeling in Marketing and Consumer Research: A review. *Int. J. Res. Mark.* **1996**, *13*, 139–161. [CrossRef]
80. Doll, W.J.; Xia, W.; Torkzadeh, G. A confirmatory factor analysis of the end-user computing satisfaction instrument. *MIS Q.* **1994**, *18*, 453–461. [CrossRef]
81. Hu, L.; Bentler, P.M. Fit indices in covariance structure modeling: Sensitivity to under parameterized model misspecification. *Psychol. Methods* **1998**, *3*, 424–453. [CrossRef]
82. Hair, J.; Sarstedt, M.; Hopkins, L.; Kuppelwieser, V.G. Partial least squares structural equation modeling (PLS-SEM) An emerging tool in business research. *Eur. Bus. Rev.* **2014**, *26*, 106–121. [CrossRef]
83. Nunnally, J.C.; Bernstein, I.H. The Assessment of Reliability. *Psychom. Theory* **1994**, *3*, 248–292.
84. Fornell, C.; Larcker, D.F. Evaluating structural equation models with unobservable variables and measurement error. *J. Mark. Res.* **1981**, *18*, 39–50. [CrossRef]
85. Henseler, J.; Ringle, C.M.; Sarstedt, M. A New Criterion for Assessing Discriminant Validity in Variance-based Structural Equation Modeling. *J. Acad. Mark. Sci.* **2015**, *43*, 115–135. [CrossRef]
86. Kline, R.B. *Priciples and Practice of Structural Equation Modeling*, 3rd ed.; Guilford Publications: New York, NY, USA, 2011.
87. MacKenzie, S.B.; Podsakoff, P.M.; Podsakoff, N.P. Construct measurement and validation procedures in MIS and behavioral research: Integrating new and existing techniques. *MIS Q.* **2011**, *35*, 293–334. [CrossRef]
88. Podsakoff, P.M.; MacKenzie, S.B.; Lee, J.Y.; Podsakoff, N.P. Common method biases in behavioral research: A critical review of the literature and recommended remedies. *J. Appl. Psychol.* **2003**, *88*, 879. [CrossRef]
89. Cook, R.D.; Weisberg, S. Residuals and influence in regression. In *Chapman and Hall*; Chapman and Hall: New York, NY, USA, 1982.
90. Bollen, K.A.; Long, J.S. (Eds.) *Testing Structural Equation Models*; Sage: Newbury Park, CA, USA, 1993.
91. Byrne, B.M. Testing for multigroup equivalence of a measuring instrument: A walk through the process. *Psicothema* **2008**, *20*, 872–882.
92. Chinn, W.W. Issues and opinion on structural equation modeling. *MIS Q.* **1998**, *22*, 7–16.
93. Cohen, J. *Statistical Power Analysis for the Behavioral Science*, 2nd ed.; Psychology Press: London, UK, 1988.
94. Falk, R.F.; Miller, N.B. *A Primer for Soft Modeling*; University of Akron Press: Akron, OH, USA, 1992.
95. Aiken, L.S.; West, S.G. *Multiple Regression: Testing and Interpreting Interactions*; Sage: Newcastle upon Tyne, UK, 1991.
96. Aguinis, H.; Beaty, J.C.; Boik, R.J.; Pierce, C.A. Effect size and power in assessing moderating effects of categorical variables using multiple regression: A 30-year review. *J. Appl. Psychol.* **2005**, *90*, 94–107. [CrossRef]
97. Scheinberg, A.; Wilson, D.C.; Rodic, L. *Solid Waste Management in the World's Cities: Water and Sanitation in the World's Cities 2010*; UN Habitat-Earthscan: Nairobi, Kenya, 2010; Available online: http://www.waste.nl/sites/waste.nl/files/product/files/swm_in_world_cities_2010.pdf (accessed on 25 March 2022).
98. Collier, J.E. *Applied Structural Equation Modeling Using AMOS: Basic to Advanced Techniques*, 1st ed.; Routledge Taylor & Francis Group: New York, NY, USA, 2020.

Article

A Techno-Economic Analysis of Sustainable Material Recovery Facilities: The Case of Al-Karak Solid Waste Sorting Plant, Jordan

Esra'a Amin Al-Athamin [1], Safwat Hemidat [2], Husam Al-Hamaiedeh [1,*], Salah H. Aljbour [3], Tayel El-Hasan [4] and Abdallah Nassour [2]

1. Civil and Environmental Engineering Department, Mutah University, Al-Karak 61710, Jordan; esraabtoush2@yahoo.com
2. Department of Waste and Resource Management, University of Rostock, 18051 Rostock, Germany; safwat.hemidat2@uni-rostock.de (S.H.); abdallah.nassour@uni-rostock.de (A.N.)
3. Chemical Engineering Department, Mutah University, Al-Karak 61710, Jordan; saljbour@MUTAH.EDU.JO
4. Chemistry Department, Mutah University, Al-Karak 61710, Jordan; tayel@MUTAH.EDU.JO
* Correspondence: husamh@mutah.edu.jo; Tel.: +962-797777036

Citation: Al-Athamin, E.A.; Hemidat, S.; Al-Hamaiedeh, H.; Aljbour, S.H.; El-Hasan, T.; Nassour, A. A Techno-Economic Analysis of Sustainable Material Recovery Facilities: The Case of Al-Karak Solid Waste Sorting Plant, Jordan. *Sustainability* 2021, *13*, 13043. https://doi.org/10.3390/su132313043

Academic Editor: Silvia Fiore

Received: 25 October 2021
Accepted: 23 November 2021
Published: 25 November 2021

Publisher's Note: MDPI stays neutral with regard to jurisdictional claims in published maps and institutional affiliations.

Copyright: © 2021 by the authors. Licensee MDPI, Basel, Switzerland. This article is an open access article distributed under the terms and conditions of the Creative Commons Attribution (CC BY) license (https://creativecommons.org/licenses/by/4.0/).

Abstract: Solid waste sorting facilities are constructed and operated to properly manage solid waste for both material and energy recovery. This paper investigates the possible technical and economic performance of the Al-Karak solid waste sorting plant in order to achieve financial sustainability and increase the profits that return on the plant to cover its operating costs. A standard procedure was followed to quantify and characterize the input materials of commercial solid waste by determining the recyclable materials in the sorting products. Thus, possible different equipment and material flows through the plant were proposed. An economic model was used in order to know the feasibility of the proposed options of the plant according to three economic factors, which are net present worth (NPW), return on investment (ROI), and payback period values. The results inferred that the characterization of the input materials contains a high portion of recyclable materials of paper, cardboard, plastic, and metals, which accounted for 63%. In this case, the mass of rejected waste to be landfilled was 9%. Results for the proposed options showed that the economic analysis is feasible when working loads on three and two shifts with ROI values of 4.4 and 3.5 with a payback period of the initial cost in 2 and 3 years, respectively. Working load on one shift was not feasible, which resulted in an ROI value of less than 2 and a payback period larger than 5 years. This paper recommended operating the sorting plant at a higher input feed with a working load on three shifts daily to ensure a maximum profit and to reduce the amount of commercial solid waste prior to landfilling through the concept of sorting and recycling.

Keywords: Al-Karak sorting plant; commercial solid waste; economic analysis; technical model structure; Jordan

1. Introduction

Solid waste (SW) is an environmental problem in both developed and developing countries. Solid waste management (SWM) is a challenge for municipalities in developing countries, owing to the growing amount of waste produced, the financial strain placed on municipal budgets as a result of the high costs associated with its management, and a lack of understanding of the many factors that influence the various stages of waste management and the interconnections required to enable the entire handling system to function [1].

The substantial volume increases in SW produced, as well as the qualitative changes in its composition as a result of major changes in living standards and conditions, have complicated SWM [2]. In Jordan, millions of tons of municipal solid waste (MSW) per year are generated from various sources such as agriculture, municipal, commercial, and industrial sectors, primarily due to population growth and the recent refugee influx, which have put additional strain on the country's already strained SW infrastructure. Landfilling is the primary disposal method of MSW in Jordan. Several types of landfills are located in different areas of Jordan [3–7]. These landfills are operated by Joint Service Councils (JSCs), which usually serve multiple municipalities in the same governorate with dual supervision from the ministry of environment and the ministry of municipalities [8].

The integrated solid waste management (ISWM) concept was introduced recently in Jordan. It contains the collection, sorting, composting, and incineration of medical wastes and sanitary landfills, which started to be implemented. However, recycling, reuse, and resource recovery are still at their initial stages [3].

Waste sorting is a key step in municipal solid waste management (MSWM) for materials recycling [9]. Jordan's MSW sorting and recycling processes are still in the early stages. Non-governmental organizations and other international organizations, such as the German International Cooperation Agency (GIZ), are responsible for the vast majority of recycling pilot programs in Jordan [10]. The idea of the sorting system is to be able to recycle valuable materials that can be sold to the market at competitive prices [11].

The Al-Karak sorting plant is one of these pilot projects in Al-Karak Governorate/Jordan. It was established in April 2019 in the El-Lajjun area. The plant is currently equipped with different basic kinds of SW handling equipment such as a mechanical conveyor belt, bale presses for cardboard and plastic, and a shredder for specific types of plastic waste. Additionally, it has two waste compression trucks. The plant was designed on infrastructure suitable to receive only dry waste (recyclable materials) from the targeted sources (separated at the source) that generated from commercial areas. The most important targeted materials according to the plant design are paper and cardboard, mixed types of plastics (polyethylene (PE) and polypropylene (PP), including colored type, polyethylene terephthalate (PET), and polystyrene (PS)), ferrous and non-ferrous metals, and any other recyclable material of a marketing value. Currently, there are no advanced mechanical treatment equipment such as screeners, air sifters, ballistic separators, or NIR separators being employed within the sorting plant to handle mixed recyclable materials efficiently. Re-designing the existing sorting plant to handle mixed recyclable materials can assist in increasing the profitability of the plant by increasing the capacity of the plant and ensuring a sustainable operation in the supply chain.

Techno-economic studies can assist in identifying the strengths and weaknesses in a proposed design or operation. They can assist in the reduction of unnecessary costs and investment risk. Raul et al. [12] performed a study by proposing different equipment or advanced technologies to an existing construction and demolition waste (C&DW) plant. The study assessed the feasibility of three different production rates of recycled aggregates (100 kt, 400 kt, and 600 kt). The economic analysis used the net present value (NPV) as a performance indicator. The economic analysis involved determining the initial investment costs and the cost of equipment. In addition, the operational costs, namely energy costs, labor costs, maintenance costs, water consumption, waste disposal costs, and insurance costs, were also determined. Operational costs showed that labor costs and energy costs were responsible for 4 and 25% of the incurred costs, respectively. Cimpan et al. [13] conducted a techno-economic assessment of central sorting at material recovery facilities. The study considered the case of lightweight packaging waste in Germany. The researchers developed four models, each with a different capacity (size) and technical degree. They revealed the cost impact of economies of scale, as well as complimentary relationships between capacity, technology, and process efficiency. As a result, a fourfold increase in capacity resulted in a threefold increase in total capital investment and a 2.4-fold increase in annual operational expense. Volk et al. [14] assessed the mechanical recycling, chemical recycling, and sequential complementary combination of both with respect to global warming potential (GWP), cumulative energy demand (CED), carbon efficiency, and product costs. The techno-economic and environmental assessment approach considered a case study on the recycling of separately collected mixed lightweight packaging (LWP) waste in

Germany. In comparison to the baseline scenario with the state-of-the-art mechanical recycling in Germany, combined mechanical and chemical recycling of LWP waste possessed significant savings potential in terms of GWP, CED, cost, and a 16% better carbon efficiency. Larrain et al. [15] investigated the economic feasibility of mechanical recycling for plastic waste. The findings revealed that the economic incentives for recycling plastic packaging are mostly determined by the product price and yield. In a scenario with steadily rising oil prices, the most profitable plastic fraction to be recycled is polystyrene, which has an internal rate of return of 14 percent, whereas the least profitable feed is a mixed polyolefin fraction, which has a negative internal rate of return. Mechanical recycling is not viable if no policy changes are enforced by governments, assuming a discount rate of 15% over a 15-year period.

The main objective of this study as to investigate the possible technical improvements that can be installed at the existing sorting plant in Al-Karak city, Jordan, in order to achieve financial sustainability that returns on the plant to cover its operating costs and to achieve an environmental aim by reducing the amount of commercial solid waste (CSW) that has to be disposed of in the landfill. First, a standard procedure was followed to characterize CSW generated in Al-Karak. Second, an economic analysis was used to evaluate the feasibility of the proposed options model of the sorting plant, which relies on present worth (PW), return on investment (ROI), and payback period values.

2. Materials and Methods

2.1. Study Area

The study area was restricted to Greater Al-Karak Municipality (Qasabah). According to the Jordan National Census, the population of Greater Al-Karak Municipality in 2019 was 112,060 [16].

The municipality and the JSC in Al-Karak are daily responsible for the cleaning, collection, and disposal of SW generated from households and commercial areas and management of the El-Lajjun landfill. MSW in Greater Al-Karak Municipality is collected from 15 districts. These districts are: Al-Karak city, Zaid Bin Al-Harithah, Al-Hiwiyah and Al-Talajah, Wadi Al-Karak, Al-Marj, Al-Thaniyah, Zahhoom, Al-Jadidah, Manshiat Abu Hammour, Al-Waysiyyah and Rakin, Al-Ghwair, Batir, Al-Adnanyah, Adr, and Al-Shihabiyah. A recent study carried out by Al-Hajaya et al. indicated that the average daily production rate of MSW in Greater Al-Karak Municipality is 61.50 tons/day [7].

The Al-Karak sorting plant is located opposite the El-Lajjun landfill site. The current daily MSW input feed to the landfill is between 200 and 250 tons/day, and the rate of CSW from the MSW is about 20% [17].

2.2. Input Materials Definition

CSW is defined as all types of SW generated by for-profit or non-profit retail stores, offices, restaurants, warehouses, education sectors, entertainment sectors, and other non-manufacturing activities, excluding residential, industrial, construction, and institutional waste or other MSW generated by home-based businesses. CSW contains a high portion of recyclable materials [18]. In this study, CSW fed to the plant was collected from different commercial areas in order to investigate the potential of utilizing CSW as recyclable materials.

2.3. CSW Collection Process

The parent population for a CW analysis campaign is the whole quantity of commercial waste, which may be sampled from and subsequently analyzed. This may encompass the whole area of a municipality or a defined part of a municipality.

Based on the "Methodology for the Analysis of Solid Waste" of the European Commission [19], a number of criteria must be applied in conducting the sampling. The total number of samplings required must consider the variation (heterogeneity) of the waste, expressed by the natural variation coefficient, as well as the number of samplings required to obtain the desired accuracy of results. In the case that the variation coefficient is unknown due to the unavailability of results from past waste analyses, it is recommended to have a sample size of 100 m^3 to carry out the waste analysis. However, in our case, the above-mentioned procedure could not be applied since the variation coefficient was unknown due to the unavailability of results from past waste analyses. In addition, it was impossible to collect 100 m^3 of CW due to limited logistics.

To avoid a low level of accuracy in the results, all the districts in Greater Al-Karak Municipality responsible for CW generations were considered for collecting the waste. The study included every sector in the commercial areas that consist of malls, supermarkets, restaurants, and garden centers. These commercial areas are supposed to supply the sorting plant of SW. Furthermore, CSW collected from offices and the educational sectors at the Military Wing in Mutah University was carried out through a collaboration between Greater Al-Karak Municipality and Mutah University. Table 1 indicates the targeted commercial areas in this study.

Table 1. Targeted commercial areas in Al-Karak city (Input materials).

Raw Materials Area	Targeted Area	Area Code	Sector
Greater Al-Karak Municipality	Mutah University/Military Wing	A1	The military offices Educational halls of students
	Mutah	A2	Military Consumer Corporation
	Al-Adnanya	A3	Malls
	Al-Karak city	A4	Restaurants Gardens Supermarkets
	Al-Marj	A5	Supermarkets
	Al-Thaniyah	A6	Supermarkets
	Zahhoom	A7	Supermarkets

It is worth mentioning that the targeted areas considered in this study for CW collection are responsible for approximately 48% of MSW generated in Greater Al-Karak Municipality [7]. The selected targeted areas being characterized by a high MSW contribution within the municipality will render an adequate and representative CW analysis result.

The study involved awareness programs for the staff responsible for SW sorting at the source in the selected targeted areas. Special bins are also distributed to collect the sorted SW. Table 2 shows the scheduled time for the source-separated CSW collection process.

Table 2. Schedule time for source-separated CSW collection process.

Days	Date	Sources	Total Weight (kg)	Empty (kg)	Net Weight (kg)
Sunday	5 July 2020	A1 + A2 + A3	10,835	10,455	380
Monday	6 July 2020	A1 + A2 + A3	10,820	10,455	365
Tuesday	7 July 2020	A1 + A2 + A3	10,775	10,455	320
Wednesday	8 July 2020	A1 + A2 + A3	10,670	10,455	215
Thursday	9 July 2020	A4 + A5 + A6 + A7	10,810	10,455	355
Friday	10 July 2020	A4 + A5 + A6 + A7	10,782	10,455	327
Saturday	11 July 2020	A4 + A5 + A6 + A7	10,735	10,455	280
Total					2242 kg

According to the time schedule, all of the targeted areas were numbered by special codes based on the farthest area from the sorting plant to the closest area, then the source-separated CSW collection process was divided into two shifts during seven consecutive days in July 2020, based on the locations of the targeted areas and the business hours (especially in the Military Wing at Mutah University). The first shift of the collection process involved the targeted areas (A1, A2, and A3), respectively, at 10:00 a.m. for a period of four days starting from Sunday to Wednesday. Then, the second shift, which targeted the areas (A4, A5, A6, and A7), respectively, at 2:00 p.m. for a period of three days starting from Thursday to Saturday. The sorting truck started the collection process of the source-separated SW from the bins and cardboard cages, then transported them to the plant in the El-Lajjun area. The weight of the collection truck was carried out through a weighbridge balance, then the collection truck unloaded the CSW at the reception site. At the end of the source-separated collection process, a total amount of 2242 kg was collected from the targeted areas.

2.4. CSW Sorting Process

After the delivery of about 2242 kg of CSW, the waste bags were opened by knives and sorted on site manually to separate the waste into fractions. The manual sorting was carried out on the floor of the reception site of the plant without using the conveyor belt. The sorting process was conducted by eight workers (two days, four working hours per day). The weight of each sorted fraction of waste was measured by using an electronic platform balance. The cardboard fraction was first pressed into a bale, then the weight of the bale was measured.

2.5. CSW Categorization

Waste characterization is an essential component of a waste analysis or the determination of waste composition. Waste characterization was carried out on the basis of the "Methodology for the Analysis of Solid Waste" of the European Commission [18]. The CSW compositions were divided into ten categories. As shown in Table 3, CSW was subdivided into two primary categories (organic fractions and inorganic fractions). The organic fraction consists of materials generated from garden centers and restaurants such as food waste and green waste. The inorganic fraction consists of plastic, cardboard, paper, and metals. Plastics were further classified into four types such as polyethylene terephthalate (PET), polypropylene (PP), polyethylene (PE), and polystyrene (PS). Metals were divided into ferrous and non-ferrous (aluminum), while the rest of the waste consisted of textiles, glass, wood, hygienic products, and compound materials.

Table 3. Sorting fractions of waste samples (manual sorting analysis).

No.	Sorting Fractions	Examples
1	PET	Lightweight plastic packaging foods and beverages, especially convenience-sized soft drinks, juices, and water.
2	PP	Containers of milk, motor oil, shampoos, soap bottles, detergents, and bleaches.
3	PE	Plastic bags, freezer bags, covering and packaging film.
4	PS	Yogurt pots, vending cups, egg cartons, and plastic cutlery.
5	Cardboard	Carton, boxes.
6	Paper	Newspaper, magazine, office paper, writing paper.
7	Ferrous metal	Tin cans and bi-metal cans, magnetic drink and food cans.
8	Non-ferrous metal	Non-magnetic drink and food cans, other non-magnetic metals such as aluminum cans.
9	Organic	Food waste and garden waste (weeds, trees, and shrub cuttings).
10	Other waste (residual)	Textiles, glass, wood, hygienic products, composite materials, and other impure materials.

The percentage of each waste category of the sorted output products were calculated according to the following formula:

$$\% \text{ of sorted output products} = \frac{M_{output}}{M_{inpt\ feed}} \times 100 \qquad (1)$$

where:

M_{output}: the mass of an output product;
$M_{input\ feed}$: the total feed input including residues quantities.

2.6. Economic Model

The information obtained from the sorting and categorization process is utilized to assess different technical scenarios for the sorting plant to operate. The different technical scenarios involve the installation of different equipment at the existing sorting plant in Al-Karak. Each scenario is economically analyzed and assessed. The economic feasibility comprised determining the capital investment costs, operating costs, and revenues of the plant's output recyclable products. The capital costs included the installed costs of the new equipment to be added to the existing sorting plant.

The operating costs included labor salaries, diesel consumption costs, electricity consumption costs, insurance costs, and maintenance and depreciation costs. The annual maintenance and depreciation costs are assumed to be 3.5% of the total capital cost. The annual insurance cost is assumed to be 0.25% of the total capital cost.

Transportation cost was not included in the operating cost. The recyclable products from the sorting plant are sold through bidding, in which bidders are committed to transferring the products from the plant. The disposal cost was not included in the operating cost because the process of waste disposal waste is the responsibility of Greater Al-Karak Municipality.

The revenues are estimated according to the current Jordanian market prices in Jordanian dinar (JOD). The selling price for either paper or cardboard is 35 JOD per ton. Ferrous metals will be treated as steel with a selling price of 65 JOD per ton. Non-ferrous metals are treated as aluminum with a selling price of 600 JOD per ton. The selling prices of PET, PP, and PE are 80, 250, and 130 JOD per ton, respectively.

The economic feasibility was assessed by calculating three economic indicators, namely: net present worth (NPW) of the project, return on investment (ROI), and payback period. The economic indicators used in the assessment process were calculated according to the following formulae:

$$NPW = \text{Initial Investment} + \sum PW \qquad (2)$$

$$PW = F \times \frac{1}{(1+i)^N} \qquad (3)$$

$$ROI = \frac{\text{Sum of all profits}}{\text{Initial Investment}} \qquad (4)$$

where:

F is the annual revenues = (Annual sales) − (annual expenses), i is the interest rate (10%), and N is the operating year. A lifetime of 15 years is assumed for the plant.

The payback period is the amount of time it takes to recover an investment's initial cost. It is the period of years it would take to recover a project's original expenditure [20].

3. Results and Discussions

3.1. CSW Characterization

CSW was classified into ten categories. Figure 1 shows the percentage of each waste fraction in the CSW. Table 4 shows the weight of each waste fraction in the CSW.

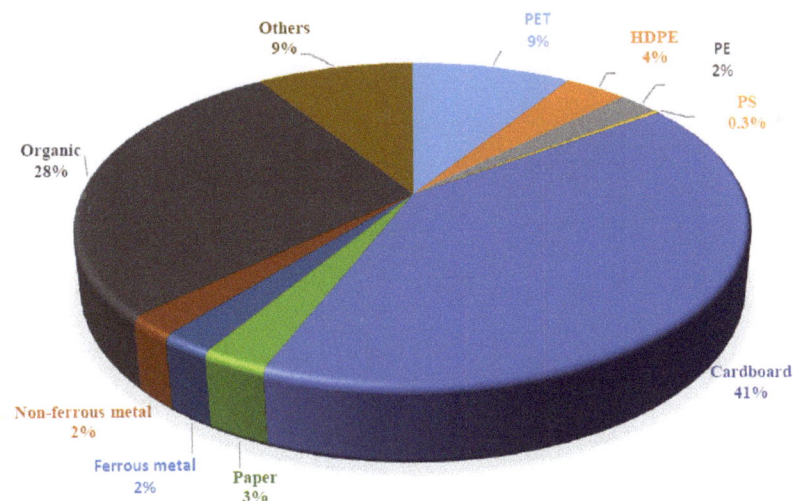

Figure 1. Waste fractions in the main input CSW from Al-Karak city (terms used in Table 3).

Table 4. Weight of waste fractions in the main input CSW from Al-Karak city.

No.	Composition	Weight (kg)	Fractions (%)
1	PET	206	9
2	PE	80	4
3	PP	52	2
4	PS	6	0.3
5	Cardboard	915	41
6	Paper	61	3
7	Ferrous metal	48	2
8	Non-ferrous metal	46	2
9	Organic	627	28
10	Other waste	201	9
	Total	2242	100

An inspection of the waste fraction results indicates that the overall CSW composition contained three main parts, namely cardboard (41%), organic matter (28%), and plastics (15%). The remainder accounted for 16% of the total and contained 4% metals, 3% paper, and 9% other waste. The CSW contained a high portion of recyclable materials such as paper, cardboard, plastic, and metals, which accounted for 63%. These findings are in agreement with the results reported in the literature. For example, results of the federal ministry for the environment in Germany indicated the presence of around 52% of recyclable materials in CSW in Germany [21]. Another study indicated 60% of potential recyclables in CSW in Germany [22].

The organic fraction can be utilized for compost production by aerobic digestion in the compost plant, which is constructed beside the sorting plant. Pilot studies indicated that high-quality compost can be produced when the source-separated organics (food and green) are utilized [4]. Figure 2 shows the material flow of CSW from Al-Karak city.

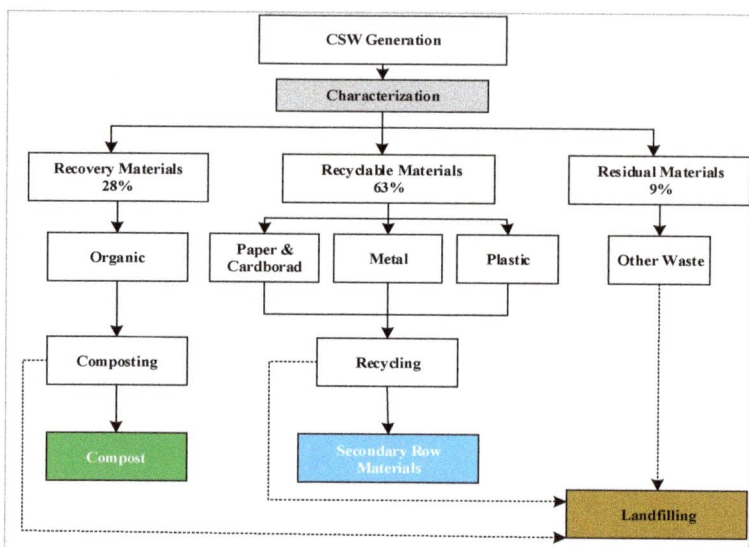

Figure 2. Material flow of CSW compositions from Al-Karak city.

It can be seen that the waste recycling and recovery are 63% and 28%, respectively, while the expected waste to be sent to landfills is 9% of the generated CSW. From environmental and economic perspectives, waste recycling and recovery will reduce the demand for new raw materials, such as cardboard, paper, and plastic. It will reduce the amount of waste generated and thus decrease the pressure on landfills and contribute to increasing the lifetime of the landfill. In addition, it will reduce the emissions of odors, landfill gas, and leachate.

The best utilization of CWS requires technical improvements in the design of the current sorting plant. Different equipment can be installed at the existing sorting plant to enhance the recycling process to achieve maximum economic return. The next sections provide an economic analysis of the recycling process.

3.2. Potential Revenues from Waste Recyclables

The current daily MSW input feed to landfills is between 200 and 250 tons/day, and the rate of CSW from the MSW is about 20% [17]. Accordingly, an average of 45 tons/day (11,925 tons/year) of CSW is expected to be generated in Al-Karak Governorate. Assuming that the CSW fractions in Al-Karak city are the same as the CSW fractions in Al-Karak Governorate, then the potential amounts of waste recyclables can be estimated along with the expected sales (Table 5).

Table 5. Potential amounts of waste recyclables and the expected sales at Al-Karak Governorate.

Potential Recyclables	Generation Rate (tons/Year *)	Sales (JOD/Year)
PET	11,925 × 0.09 = 1073	85,860
PE	11,925 × 0.04 = 477	62,010
PP	11,925 × 0.02 = 239	59,625
Cardboard	11,925 × 0.41 = 4889	171,124
Paper	11,925 × 0.03 = 358	12,521
Ferrous metal	11,925 × 0.02 = 239	15,503
Non-ferrous metal	11,925 × 0.02 = 239	143,100
Total	7513	549,743

* Year for collection waste = 265 days.

In order to achieve these annual revenues from Al-Karak Governorate, high levels of awareness and public participation are needed. Appropriate legislation from the municipalities and financial incentives are also needed to promote public awareness with respect to separation at the source. Effective collection and sorting of CSW are expected to achieve annual sales at 549,743 JOD/year. The technical possible improvement that can be installed to the current situation of the sorting plant is to install an NIR separator and ballistic separator. Figure 3 shows the flow diagram of the new technical structure of the sorting process.

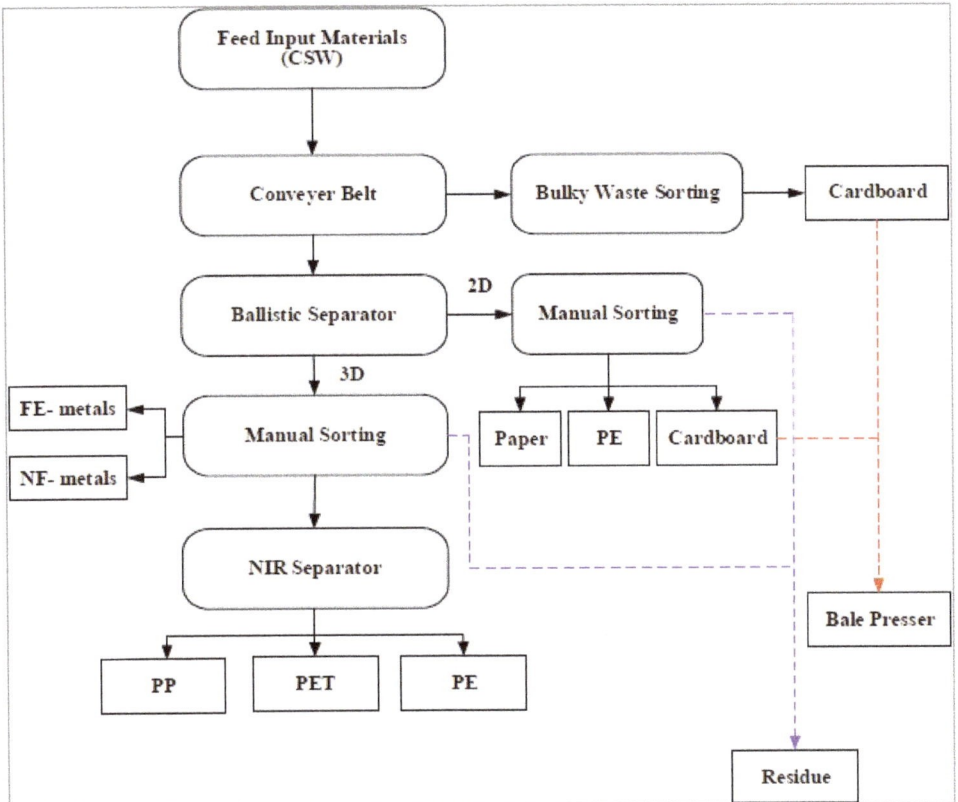

Figure 3. Flow diagram of the new technical structure of the sorting process.

As can be seen in Figure 3, after bulky waste sorting, the input feed (CSW) will be directed into a ballistic separator where materials are separated based on their shapes into 2D and 3D materials. The 2D feed is transferred to the manual sorting. Here, the components will mainly contain PE, paper, and cardboard, while the 3D feed is transferred to the manual sorting firstly to separate metals, then to the NIR separator in order to separate different kinds of plastics efficiently such as PET, PP, and PE.

3.3. Capital Investment Costs

The current sorting plant has a bale presser, one conveyor belt, plastics shredder, forklift, and two waste collection trucks, all operating at high efficiency. The sorting plant has a hanger constructed on land owned by the municipality. Therefore, the new proposed design of the sorting plant requires only the purchase and installation of the new equipment along with their auxiliaries as capital costs. The total capital cost for the new design is detailed in Table 6.

Table 6. Total capital cost for the new design.

Equipment	Cost in (JOD) [22]
Ballistic separator	127,000
NIR @2.0 m belt width	185,000
Conveyor belt	12,000
Air Compressor for NIR	70,000
Total Capital Costs	394,000

3.4. Annual Operating Cost

3.4.1. Worker and Personnel Salaries

The total number of employees at the sorting plant is 33, working three shifts a day. The employees are 6 drivers, 24 workers, and 3 officers. The annual costs due to salaries are illustrated in Table 7 [23].

Table 7. Labor and personnel salaries.

Work Force	Number	Cost per Year (JOD)
Drivers	6	23,760
Workers	24	95,040
Officers	3	15,840
Total	33	134,640

3.4.2. Utility Costs

The utility costs comprised basically the cost of fuel consumed by the waste collection trucks and the cost of electricity consumed by the equipment. Each waste collection truck has a capacity for diesel fuel of 100 L. The 100 L will be consumed in 2 days; accordingly, the number of filling times in the year will be:

$$\text{Filling times per year} = \frac{\text{working days}}{\text{Diesel consumption days}} = \frac{264}{2} = 132 \text{ times/truck/year}$$

The cost of 100 L diesel is 50 JOD; thus, the total diesel consumption per year will be:

$$132 \text{ times/truck/year} \times 2 \text{ trucks} \times 50 \frac{\text{JOD}}{\text{Filling}} = 13,200 \text{ JOD/year}$$

Electricity is supplied to the Al-Karak sorting plant through the Electricity Distribution Company (EDCO). The electrical consumption category is considered as small industries according to the EDCO [24]. Electricity costs for this category are shown in Table 8.

Table 8. Electricity costs for the category "small industries".

Category kWh	Costs JOD/kWh
1–2000	120
Above 2000	175

The proposed model requires operating the plant for 24 h through three working shifts. The total electricity consumption and costs are shown in Table 9 [23].

Table 9. Electricity consumption Costs.

Equipment	Average Power kWh	Energy per Month kWh	Energy per Year kWh
Two conveyor belts	5	5280	63,360
Bale press for cardboard	18	9504	114,048
Bale press for plastic	35	18,480	221,760
Plastic shredder	45	23,760	285,120
Ballistic separator	20	10,560	126,720
NIR separator	5	2640	31,680
Air Compressor for NIR	30	15,840	190,080
Total consumption		86,064	1,032,768
		Total	90,367 JOD/year

3.4.3. Maintenance and Depreciation Costs

Maintenance and depreciation costs are assumed to be 3.5% of the total capital cost.

$$\text{Maintenance costs} = 0.035 \times 394{,}000 = 13{,}790 \text{ JOD/year}$$

$$\text{Depreciation costs} = 0.035 \times 394{,}000 = 13{,}790 \text{ JOD/year}$$

3.4.4. Insurance Costs

Insurance costs are assumed to be 0.25% of the total capital cost.

$$\text{Insurance costs} = 0.0025 \times 394{,}000 = 985 \text{ JOD}$$

The above calculations indicate that the total annual operating cost is 266,772 JOD/year. Figure 4 shows the percentage of the contributing factors in the operating cost.

The results of the total yearly operating expenses revealed that labor and personnel costs are responsible for 50% of the yearly expenses. Electricity consumption costs accounted for 34% of the yearly expenses. Depreciation, maintenance, and fuel consumption costs were comparable to each other and contributed to 15% of the yearly expenses.

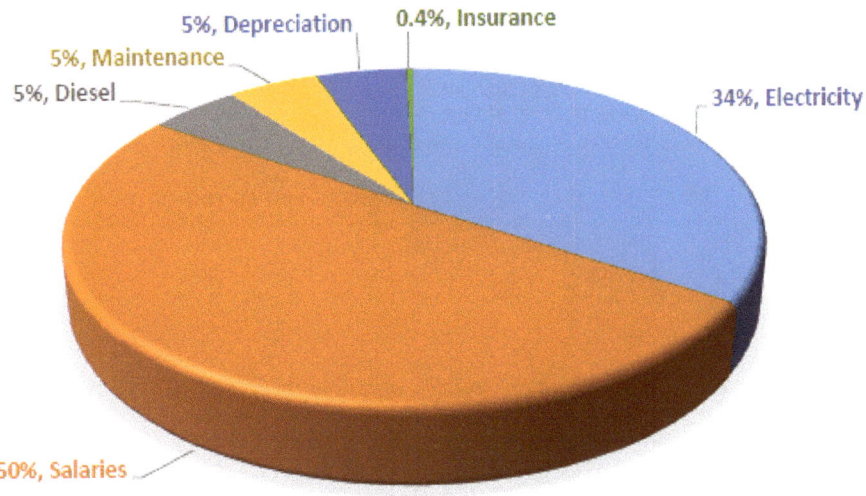

Figure 4. Percentage of operating cost components.

3.5. Economic Feasibility Analysis

To evaluate the economic feasibility of the new sorting plant design, three important economic indicators must achieve acceptable performance. Feasible investments must have a positive NPW value, an ROI of >2.0, and an acceptable payback period <5 years. Table 10 shows the cash flow calculations for the proposed sorting plant.

Table 10. Cash flow calculations for the proposed sorting plant.

Year	Capital Cost (JOD)	Sales (JOD)	Operating Costs	Revenues (JOD)	Discounting Factor @10%	PW (JOD)	Cumulative PW (JOD)
0	−394,000	0		−394,000	1	−394,000	−394,000
1	0	549,743	−266,772	282,971	0.909	257,246	−136,754
2	0	549,743	−266,772	282,971	0.826	233,860	97,107
3	0	549,743	−266,772	282,971	0.751	212,600	309,707
4	0	549,743	−266,772	282,971	0.683	193,273	502,980
5	0	549,743	−266,772	282,971	0.621	175,703	678,683
6	0	549,743	−266,772	282,971	0.564	159,730	838,412
7	0	549,743	−266,772	282,971	0.513	145,209	983,621
8	0	549,743	−266,772	282,971	0.467	132,008	1,115,629
9	0	549,743	−266,772	282,971	0.424	120,007	1,235,637
10	0	549,743	−266,772	282,971	0.386	109,098	1,344,734
11	0	549,743	−266,772	282,971	0.350	99,180	1,443,914
12	0	549,743	−266,772	282,971	0.319	90,163	1,534,077
13	0	549,743	−266,772	282,971	0.290	81,967	1,616,044
14	0	549,743	−266,772	282,971	0.263	74,515	1,690,559
15	0	549,743	−266,772	282,971	0.239	67,741	1,758,300

The cash flow calculations indicate a potential NPW of 1,758,300 JOD. This indicates a rate of investment of 4.4. The payback period is less than 2 years.

The economic feasibility was re-assessed for cases when the plant is operated under one and two working shifts.

To operate the plant with two shifts, it is assumed that a total number of 22 personnel is needed and 2/3 the amount of diesel is needed. In this case, the total operating expenses in the two shifts = 187,370 JOD/year.

To operate the plant with one shift, it is assumed that a total number of 11 personnel is needed and 1/3 the amount of diesel is needed. In this case, the total operating expenses in the two shifts = 107,967 JOD/year.

The NPW, ROI, and the payback period for the new scenarios are shown in Table 11.

Table 11. Effect of the number of working shifts on the economic indicators.

Shift No.	NPW	ROI	Payback Period
3 shifts	1,758,300	4.4	2
2 shifts	982,016	3.5	3
1 shift	185,378	1.5	8

The results shown in Table 10 indicate that operating the sorting plant using two working shifts is economically feasible with an ROI value of 3.5 and a payback period of the initial cost in three years. However, operating the sorting plant using one working shift is economically not feasible, as the resulting ROI was less than 2 with a payback period of 8 years.

4. Conclusions

- The sorting analysis of the input materials inferred that the CSW was collected from seven targeted commercial areas and contains a high portion of recyclable materials

of paper, cardboard, plastic, and metals, which accounted for 63%. In this case of the recovery of CSW materials from MSW, the mass of rejected waste to be landfilled was 9%.
- The technical possible improvement that was installed in the sorting plant was the addition of an NIR separator and a ballistic separator. Analyzing the results of these proposed options of the plant showed that.
- The economic analysis was feasible when working loads on three and two shifts with ROI values of 4.4 and 3.5, respectively, whereas working load on one shift was not feasible, which resulted in an ROI value less than 2 and a payback period larger than 5 years.

Author Contributions: Conceptualization, A.N. and H.A.-H.; methodology, E.A.A.-A.; H.A.-H.; software E.A.A.-A., S.H.A.; validation, T.E.-H. and S.H.; formal analysis, E.A.A.-A.; H.A.-H.; investigation, S.H.A.; resources, E.A.A.-A.; data curation, H.A.-H. and S.H.; writing—original draft preparation, E.A.A.-A.; writing—review and editing, H.A.-H., S.H.A. and T.E.-H.; supervision, H.A.-H.; project administration, H.A.-H.; funding acquisition, T.E.-H. All authors have read and agreed to the published version of the manuscript.

Funding: This research was funded by Deutsche Gesellschaft für Internationale Zusammenarbeit (GIZ) Grant number 84190076 in the framework of the Waste to Positive Energy (Wt(P)E) project, Jordan.

Institutional Review Board Statement: Not applicable.

Informed Consent Statement: Not applicable.

Data Availability Statement: The data reported in this study does not represent a publicly archived datasets.

Acknowledgments: This research is accomplished as a contribution to the Waste to Positive Energy (Wt(P)E) project in Jordan. The project is supported and funded by GIZ and with the participation of three German universities (University of Rostock (Rostock, Germany), TU Hamburg (Hamburg, Germany) and TU Dresden (Dresden, Germany)) and four Jordanian universities (Jordan University of Science and Technology (Irbid, Jordan), Mutah University (Karak, Jordan), University of Jordan (Amman, Jordan), and German-Jordanian University (Madaba, Jordan)).

Conflicts of Interest: The authors declare no conflict of interest.

References

1. Guerrero, L.A.; Maas, G.; Hogland, W. Solid waste management challenges for cities in developing countries. *Waste Manag.* **2013**, *33*, 220–232. [CrossRef] [PubMed]
2. Ikhlayel, M.; Higano, Y.; Yabar, H.; Mizunoya, T. Introducing an integrated municipal solid waste management system: Assessment in Jordan. *J. Sustain. Dev.* **2016**, *9*, 43. [CrossRef]
3. Hemidat, S. Feasibility Assessment of Waste Management and Treatment in Jordan. Ph.D. Thesis, Universität Rostock, Rostock, Germany, 2019.
4. Al-Nawaiseh, A.R.; Aljbour, S.H.; Al-Hamaiedeh, H.; El-Hasan, T.; Hemidat, S.; Nassour, A. Composting of Organic Waste: A Sustainable Alternative Solution for Solid Waste Management in Jordan. *Jordan J. Civ. Eng.* **2021**, *15*, 363–377.
5. Aljbour, S.H.; El-Hasan, T.; Al-Hamiedeh, H.; Hayek, B.; Abu-Samhadaneh, K. Anaerobic co-digestion of domestic sewage sludge and food waste for biogas production: A decentralized integrated management of sludge in Jordan. *J. Chem. Technol. Metall.* **2021**, *56*, 1030–1038.
6. Aljbour, S.H.; Al-Hamaiedeh, H.; El-Hasan, T.; Hayek, B.O.; Abu-Samhadaneh, K.; Al-Momany, S.; Aburawaa, A. Anaerobic co-digestion of domestic sewage sludge with food waste: Incorporating food waste as a co-substrate under semi-continuous operation. *J. Ecol. Eng.* **2021**, *22*, 1–10. [CrossRef]
7. Al-Hajaya, M.; Aljbour, S.H.; Al-Hamaiedeh, H.; Abuzaid, M.; El-Hasan, T.; Hemidat, S.; Nassour, A. Investigation of Energy Recovery from Municipal Solid Waste: A Case Study of Al-Karak City/Jordan. *Civ. Environ. Eng.* **2021**. [CrossRef]
8. Aljaradin, M.; Persson, K.M. Solid waste management in Jordan. *Int. J. Acad. Res. Bus. Soc. Sci.* **2014**, *4*, 138–150. [CrossRef]
9. Gundupalli, S.P.; Hait, S.; Thakur, A. A review on automated sorting of source-separated municipal solid waste for recycling. *Waste Manag.* **2017**, *60*, 56–74. [CrossRef] [PubMed]
10. Jordan_Green_Building_Council. *Waste Sorting Informative Booklet*; Jordan Green Building Council: Amman, Jordan, 2016.

11. Rives, J.; Rieradevall, J.; Gabarrell, X. LCA comparison of container systems in municipal solid waste management. *Waste Manag.* **2010**, *30*, 949–957. [CrossRef] [PubMed]
12. Neto, R.O.; Gastineau, P.; Cazacliu, B.G.; Le Guen, L.; Paranhos, R.S.; Petter, C.O. An economic analysis of the processing technologies in CDW recycling platforms. *Waste Manag.* **2017**, *60*, 277–289. [CrossRef]
13. Cimpan, C.; Maul, A.; Wenzel, H.; Pretz, T. Techno-economic assessment of central sorting at material recovery facilities—The case of lightweight packaging waste. *J. Clean. Prod.* **2016**, *112*, 4387–4397. [CrossRef]
14. Volk, R.; Stallkamp, C.; Steins, J.J.; Yogish, S.P.; Müller, R.C.; Stapf, D.; Schultmann, F. Techno-economic assessment and comparison of different plastic recycling pathways: A German case study. *J. Ind. Ecol.* **2021**, *25*, 1318–1337. [CrossRef]
15. Larrain, M.; Van Passel, S.; Thomassen, G.; Van Gorp, B.; Nhu, T.T.; Huysveld, S.; Van Geem, K.M.; De Meester, S.; Billen, P. Techno-economic assessment of mechanical recycling of challenging post-consumer plastic packaging waste, Resources. *Conserv. Recycl.* **2021**, *170*, 105607. [CrossRef]
16. Department of statistics. *Estimated Population of the Kingdom by Municipality and Sex, at End-Year 2019*; Department of Statisitics: Amman, Jordan, 2019.
17. *Annual Report 2018: Solid and Commercial Waste*; Karak_Greater_Municipality: Karak, Jordan, 2018.
18. UBA, *Quantity, Management and Aspects of Resource Preservation of Commercial Solid Waste*; Federal Environment Agency (Umweltbundesamt): Dessau-Roßlau, Germany, 2011.
19. SWA-Tool Consortium. *Methodology for the Analysis of Solid Waste (SWA-Tool) User Version*; European Commission: Luxembourg, 2004.
20. Sullivan, W.; Wicks, E.; Koelling, C. *Engineering Economy*, 16th ed.; Pearson: Upper Saddle River, NJ, USA, 2015.
21. UBA. *Commercial Municipal Waste*; Federal Ministry for the Environment (Umweltbundesamt): Dessau-Roßlau, Germany, 2013.
22. Nelles, M.; Gruenes, J.; Morscheck, G. Waste management in Germany–development to a sustainable circular economy? *Procedia Environ. Sci.* **2016**, *35*, 6–14. [CrossRef]
23. *Annual Report 2019: Al-Kark Solid Waste Sorting Plant—Financial Report*; Karak_Greater_Municipality: Karak, Jordan, 2019.
24. EDCO, Electricity Distribution Company, Jordan, Annual Report 2020. Available online: https://www.edco.jo/index.php/en/ (accessed on 12 November 2021).

Article

Waste to Hydrogen: Elaboration of Hydroreactive Materials from Magnesium-Aluminum Scrap

Olesya A. Buryakovskaya [1,*], Anna I. Kurbatova [2], Mikhail S. Vlaskin [1], George E. Valyano [1], Anatoly V. Grigorenko [1], Grayr N. Ambaryan [1] and Aleksandr O. Dudoladov [1]

1 Laboratory of Energy Storage Substances, Joint Institute for High Temperatures of the Russian Academy of Sciences, Izhorskaya Street, 13, Build. 2, 125412 Moscow, Russia; vlaskin@inbox.ru (M.S.V.); gvalyano@yandex.ru (G.E.V.); presley1@mail.ru (A.V.G.); ambaryan1991@gmail.com (G.N.A.); nerfangorn@gmail.com (A.O.D.)
2 Department of Environmental Safety and Product Quality Management, Institute of Environmental Engineering, Peoples' Friendship University of Russia (RUDN University), 6 Miklukho-Maklaya Street, 117198 Moscow, Russia; kurbatova-ai@rudn.ru
* Correspondence: osminojishe@yandex.ru

Citation: Buryakovskaya, O.A.; Kurbatova, A.I.; Vlaskin, M.S.; Valyano, G.E.; Grigorenko, A.V.; Ambaryan, G.N.; Dudoladov, A.O. Waste to Hydrogen: Elaboration of Hydroreactive Materials from Magnesium-Aluminum Scrap. *Sustainability* **2022**, *14*, 4496. https://doi.org/10.3390/su14084496

Academic Editor: Giovanni Esposito

Received: 19 February 2022
Accepted: 7 April 2022
Published: 10 April 2022

Publisher's Note: MDPI stays neutral with regard to jurisdictional claims in published maps and institutional affiliations.

Copyright: © 2022 by the authors. Licensee MDPI, Basel, Switzerland. This article is an open access article distributed under the terms and conditions of the Creative Commons Attribution (CC BY) license (https://creativecommons.org/licenses/by/4.0/).

Abstract: Ball-milled hydroreactive powders of Mg-Al scrap with 20 wt.% additive (Wood's alloy, KCl, and their mixture) and with no additives were manufactured. Their hydrogen yields and reaction rates in a 3.5 wt.% NaCl aqueous solution at 15–35 °C were compared. In the beginning of the reaction, samples with KCl (20 wt.%) and Wood's alloy (10 wt.%) with KCl (10 wt.%) provided the highest and second-highest reaction rates, respectively. However, their hydrogen yields after 4 h were correspondingly the lowest and second-lowest percentages—(45.6 ± 4.4)% and (56.0 ± 1.2)% at 35 °C. At the same temperature, samples with 20 wt.% Wood's alloy and with no additives demonstrated the highest hydrogen yields of (73.5 ± 10.0)% and (70.6 ± 2.5)%, correspondingly, while their respective maximum reaction rates were the lowest and second-lowest. The variations in reaction kinetics for the powders can be explained by the difference in their particle sizes (apparently affecting specific surface area), the crystal lattice defects accumulated during ball milling, favoring pitting corrosion, the morphology of the solid reaction product covering the particles, and the contradicting effects from the potential formation of reaction-enhancing microgalvanic cells intended to induce anodic dissolution of Mg in conductive media and reaction-hindering crystal-grain-screening compounds of the alloy and metal scrap components.

Keywords: magnesium-aluminum scrap; ball milling; potassium chloride; Wood's alloy; hydroreactive powders; simulated sea water; hydrogen production

1. Introduction

In recent years, waste management has become one of the mainstream environmental topics for scientists, entrepreneurs, and politicians worldwide. According to the report 'What a Waste 2.0: A Global Snapshot of Solid Waste Management to 2050', the global annual waste generation is expected to rise from 2.01 billion tons in 2016 to 3.5 billion tons in 2050. Today, in high-income countries, about 40% of waste is disposed of in landfills, another 40% undergoes recycling, and energy is recovered from 20% of waste. In low-income countries, over 90% of waste is either burned or openly dumped, and only 4% of waste is recycled. In Russia, almost 96% of solid waste—with metals contributing 4% to the total amount—is buried in landfills; municipal solid waste is collected as mixed aggregations without any sorting efforts and has a moisture content between 30–40% [1,2].

Some sorts of metal waste can be effectively recycled on the condition that its separate collection and proper disposal are ensured. In study [3], the sustainability of different approaches to collecting the metal fraction of household waste was examined. The results confirmed that, in terms of greenhouse gas emissions, separate collection and recycling of the municipal solid waste metallic fraction at residential properties represented the preferable option compared to a scenario with no source sorting and incineration of everything. However, such measures of efficient metal waste handling are still not implemented widely. Not only households, but many small metal mechanical enterprises as well, negatively affect the environment by inadequate disposal of manufacturing and raw material wastes generated in their production processes [4].

Metal processing scrap of light metals, such as magnesium, aluminum, and their alloys, can be used for in situ hydrogen production by their oxidation in aqueous media [5]. The typical solid reaction products—$Mg(OH)_2$, $Al(OH)_3$, and $AlOOH$—do not produce any negative environmental impacts and can be used as raw stuff for the manufacture of various valuable materials. For example, MgO-based composites are currently attracting attention as a durable concrete for sustainable and energy-efficient building design [6]. Al_2O_3-based materials are used for the manufacture of heat-resistant ceramics, porous catalyst carriers, synthetic sapphire for microelectronics, etc. [7]. Additionally, the said solid reaction by-products can be returned to the metal production cycles and, thereby, eliminate the process stages of mining, separation of aluminum or magnesium compounds from their ores, and disposal of mud that is produced in large amounts [8]. If a scenario of secondary aluminum or magnesium production by melting, refining, and casting is preferred, some amount of magnesium- and aluminum-based scrap can be used to produce hydrogen that, in turn, can be further utilized to produce energy for the furnaces of the secondary metal industry itself [9].

Today, large amounts of Mg waste in the form of chips and discards are generated from the machining of sheets and castings [10]. Up to 30% of Mg is lost as scrap during manufacturing [11]. Extensive mechanical processing is required for the production of magnesium-based components for the automobile and aerospace industries (engine blocks, oil pans, transmission housings, seat frames, etc.) [12]. Widely used construction alloys for cyclically loaded structural applications include the NZK (Mg-Nd-Zn-Zr), AZ91D, GW103, and AM-SC grades [13]. ML10 and ML5 casting alloys have elemental compositions generally corresponding to NZK and AZ91D respectively. ML10 is applied for the loaded components of engines, instruments, and equipment requiring high levels of leak tightness. ML5 is used for manufacturing components of plane wings, chassis, and control elements, as well as for large shaped castings, such as fan housings (290 kg), compressor housings (720 kg), helicopter gearbox housings (340 kg), and other complex contoured parts [14]. Thus, magnesium-based casting alloys represent materials for which processing generates plenty of waste chips, shavings, and cuts, and this scrap should be effectively utilized.

Under normal conditions, the reaction between aluminum or magnesium and water barely proceeds because of a poorly permeable oxide film on the surfaces of these metals and the formation of a dense layer of reaction products thereon. Therefore, higher temperatures or special activation methods must be employed to carry out the reaction. Today, many studies are known on hydrogen production from aluminum or magnesium scrap and water or aqueous solutions. In pioneer studies [15–18], low-grade magnesium scrap in bulk form has been converted into hydrogen using titanium platinum-coated or stainless steel nets as a catalyst (in [15], as a grinding surface as well) and either natural sea water or a 3.5 wt.% NaCl aqueous solution, either with citric acid or without it. In study [19], different types of aluminum waste (foil, capacitor casings) were hydrothermally treated at temperatures of 230–340 °C under the corresponding saturated steam pressure, and it was established that the reaction mechanism was similar to that for pure aluminum powder. Another widely-known method for producing hydrogen from aluminum is the implementation of alkali solutions, most commonly KOH and NaOH [20–22], under moderate (below 100 °C) and hydrothermal [23] conditions.

A promising approach for obtaining hydrogen from aluminum or magnesium oxidation in aqueous media at moderate conditions without the implementation of acids or alkali solutions is the preparation of powder materials by ball milling [5,24]. Aluminum or magnesium in the form of powders or larger particles can be ball-milled either as they are or with various additives. Early studies [25,26] on magnesium ball milling have demonstrated a significant decrease in the corrosion resistance of ball-milled magnesium both in pure water and in conductive solutions. It was established as well that a ball-milled composition of Mg and Ni (10 at.%) powders substantially represented a so-called 'mechanical alloy', wherein Ni sites acted as cathodes while Mg sites acted as anodes. Due to a high electrode potential difference between Mg and Ni, intensive Mg galvanic corrosion in the conductive media (1 M KCl aqueous solution) was observed. The same 'galvanic' effects in sea water or its simulation (3.5 wt.% NaCl aqueous solution) were investigated also for a wide range of additives to magnesium: Zn, In, Co, Bi, Al, Fe, Ni, Cu, Sn, Ga, Ce, and La [27–40]. Studies [41,42] have been devoted to the investigation of the effect of other salt solutions (0.5–2.5 M NiCl$_2$ solution; NiCl$_2$ added to Marmara and Aegean sea water; 1 M CoCl$_2$, CuCl$_2$, FeCl$_3$, and MnCl$_2$ solutions) on magnesium powder obtained by the ball milling of scraps for 3–30 h. In other papers, different salts (primarily metal chlorides: AlCl$_3$, NaCl, and KCl) have been added to magnesium [43,44] or aluminum [45–48] powder, and the intensification of ball-milling effects (the formation of metal lattice structure defects, metal particle size reduction, etc.) has been demonstrated. In recent studies [49–51] devoted to the elaboration of hydroreactive powders from magnesium waste, simulated sea water (3.5 wt.% NaCl solution) has been used as aqueous media as well, and a complex ball-milling process has been implemented that included the simultaneous or chronological addition of 5 wt.% C (carbon) and Ni or AlCl$_3$ powders.

As it can be concluded from the survey of studies on hydrogen production from the oxidation of Al- and Mg-based materials in aqueous media, this subject remains relevant. A number of methods for manufacturing metal scrap-based compositions for hydrogen production have been tested. The ball milling of Mg- and Al-based disperse materials with metal or salt additives, as well as the implementation of sea water or its simulation with NaCl aqueous solutions, are still relevant. The present study is focused on the elaboration of hydroreactive materials from Mg-Al alloy scrap by its ball milling with commercially available additives in order to test their hydrogen generation properties in simulated sea water. The main aim of this study is to compare the performances (hydrogen yields and evolution rates) of different powder materials ball-milled for 4 h both with 20 wt.% additives (KCl salt, Wood's alloy (composed of low-melting-point metals), and their mixture) and without additives. The compositions and content of the additives, as well as the ball-milling duration, were selected based on the results from preceding studies.

2. Materials and Methods
2.1. Original Materials and Powder Preparation Procedure

The original materials for the preparation of hydroreactive powders included chemically pure KCl salt (National State Standard GOST 4234-77, rev. 1–2, 'VEKTON' JSC, Saint-Petersburg, Russia); pure Wood's metal alloy containing 9.7 wt.% Sn, 40.4 wt.% Pb, 9.67 wt.% Cd, and 40 wt.% Bi (Technical Specification No. 6-09-4064-87, 'Rushim' LLC, Moscow, Russia; solidification temperature—71.0 °C); and metal scrap composed mainly of waste chips of a magnesium-aluminum alloy of the ML5 grade with some amount of ML10 shavings (National State Standard GOST 2856-79) originated from mechanical processing at an aircraft manufacturing plant. The original magnesium-aluminum scrap and chips manufactured from melted Wood's alloy pellets are illustrated in Figure 1. For the preparation of the salt aqueous solution, deionized water and chemically pure NaCl salt (National State Standard GOST 4233-77, 'VEKTON' JSC, Saint-Petersburg, Russia) were used.

Prior to ball milling, in order to remove lubricating oil remaining on the surface of the scrap particles after mechanical processing, a degreasing procedure was carried out. The degreasing process included the immersion of a scrap portion into a flask filled with pure acetonitrile (Technical Specification No. 2636-092-44493179-04, 'EKOS-1' JSC, Moscow, Russia), ultrasonic cleaning for 1 h in an ultrasonic bath sonicator (PSB-2835-05; 'PSB-Gals' Ltd., Moscow, Russia), and further stirring of the 'scrap–acetonitrile suspension' by a magnetic mixer (C-MAG; 'HS 7 IKA-Werke' GmbH & Co. KG, Staufen, Germany) with a stirring bar for 1 h. Then, the 'used' acetonitrile portion was changed for a fresh one, and ultrasonic cleaning for 1 h and stirring for 1 h were repeated. The degreased metal scrap was separated from the acetonitrile and dried at ambient temperature for 24 h.

(a) (b)

Figure 1. Original materials for powder preparation: (**a**) magnesium-aluminum alloy scrap; (**b**) original Wood's metal alloy pellets and the chips manufactured thereof.

For the ball milling procedure, a 50 mL milling pot of corundum and stainless steel and 24 stainless steel milling balls of 10 mm in diameter were used. The milling pot was filled with the original materials and balls in a glove box (G-BOX-F-290; 'FUMATECH' Ltd., Novosibirsk, Russia) under pure argon (99.993%, National State Standard GOST 10157-79, 'NII KM' Ltd., Moscow, Russia). For all samples, the ball : powder mass ratio was 24:1. Ball milling was performed using a centrifugal ball mill (S 100; 'Retsch' GmbH, Haan, Germany) for 4 h at a rotational speed of 580 rpm.

2.2. Experimental Facility and Procedure

In the experiments on hydrogen evolution kinetics, an experimental facility schematically shown in Figure 2 was employed. The main functional elements of the facility included a 500 mL reactor (Simax glass) with its heating and cooling jacket connected—for experiments at 25 °C and higher temperatures—to a heater (CC-308B; 'ONE Peter Huber Kältemaschinenbau' GmbH, Offenburg, Germany) or—for experiments at temperatures below 25 °C—to a cryothermostat (LOIP FT-311-80; 'Laboratory Equipment and Instruments' Ltd., Saint-Petersburg, Russia), and the gas exhaust outlet was connected to a Drexel flask connected, in turn, to a glass vessel filled with water. The said water was ejected by incoming hydrogen into a flask placed on scales (ATL-8200d1-I; 'Acculab Sartorius Group', New York, USA) continuously transmitting data to a personal computer. The mixture in the reactor was stirred by means of a magnetic mixer (C-MAG HS 7; 'IKA-Werke' GmbH & Co. KG, Staufen, Germany) with a stirring bar. The temperatures in the reactor and in the glass vessel containing water and hydrogen were measured, respectively, with an L-type thermocouple (TP.KhK(L)-K11; 'Relsib' LLC, Novosibirsk, Russia) and a Pt100-type resistance temperature detector (TS-1288 F/11; 'Elemer' LLC, Podolsk, Russia) connected to a multichannel thermometer (TM 5103; 'Elemer' LLC, Podolsk, Russia) for data collection and storage. The atmosphere pressure was measured by an aneroid barometer (BTKSN-18; Technical Specification No. 1-099-20-85, 'UTYOS' JSC, Ulyanovsk, Russia).

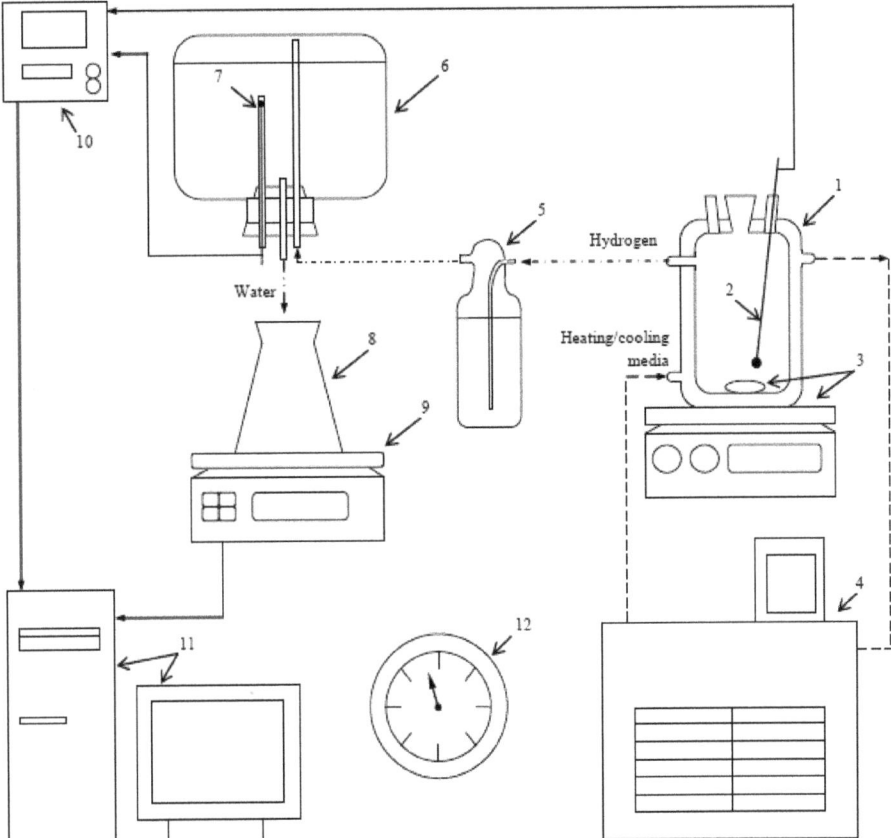

Figure 2. Experimental facility: 1—reactor, 2—thermocouple, 3—magnetic mixer and stirring bar, 4—heater or cryothermostat, 5—Drexel flask, 6—glass vessel, 7—resistance temperature detector, 8—flask, 9—scales, 10—multichannel thermometer, 11—computer, and 12—barometer.

The experimental procedure included pouring into the reactor 400 mL of 3.5 wt.% NaCl solution and heating or cooling it to the desired temperature under stirring. Upon temperature stabilization, 0.75 g of powder was loaded into the reactor. The hydrogen produced during the reaction was bubbled in a Drexel flask and then passed into a glass vessel filled with water. The water ejected by incoming hydrogen was collected in a flask placed onto scales, and its mass readings were transmitted to a computer for recording and storage. The data on the temperatures in the reactor and glass vessel were continuously recorded as well. The atmospheric pressure values were fixed at the beginning and in the end of each experiment. The duration of each experiment was 4 h. The resulting solid product was separated from the water using a Bunsen flask, Buchner funnel, paper filter, and circulating water vacuum pump (SHZ-D III; 'FAITHFUL Instrument Co.', Ltd., Hebei, China) and was dried at ambient temperature.

The data on the ejected water volume (mass), temperature in the glass vessel, and atmospheric pressure were used to calculate the hydrogen volume values at standard conditions (101,325 Pa, 0 °C) using the ideal gas law. In the present study, kinetic curves represented plots of hydrogen yield vs. time. The hydrogen yield represented the ratio of the hydrogen volume (normalized to the standard conditions) obtained in the experiments to the theoretical maximum hydrogen volume. The theoretical maximum hydrogen volume was obtained providing that the total amount of hydroreactive components in a sample was consumed by the reaction. The calculation of the hydrogen yield was performed using the following equation:

$$\alpha(\tau) = \frac{V(\tau)}{V_{\max}}, \qquad (1)$$

where $\alpha(\tau)$ is the hydrogen yield (%), $V(\tau)$ is the hydrogen volume (mL) produced in the experiment, τ is time (s), and $V_{\max}$ is the theoretical maximum hydrogen volume (mL). For each powder sample and each tested temperature point, a series of three experiments was carried out. Each of the kinetic curves was obtained by averaging the data sets obtained from the three experiments, and the standard deviations for all of them were provided.

The reaction rate constants were derived from the approximation of the kinetic curves by the Avrami–Erofeev equation for heterogeneous reactions [52]:

$$\alpha(\tau) = 1 - exp\left[-(k \cdot \tau)^n\right], \qquad (2)$$

where $\alpha(\tau)$ is the hydrogen yield (%), τ is time (s), k is the reaction rate constant (s^{-1}), and n is a nondimensional parameter depending on the nucleation rate law (derived from kinetic curve approximation as well).

The reaction rate constants were employed for the calculation of the activation energy using the Arrhenius relationship [52] in the following form:

$$\ln(k) = \ln(A) - \frac{E_a}{R} \cdot \frac{1}{T}, \qquad (3)$$

where k is the reaction rate constant (s^{-1}), E_a is the activation energy (J), R is the universal gas constant, T is temperature (K), and A is the pre-exponential factor (frequency factor).

2.3. Methods and Equipment for Sample Analyzing

For the hydroreactive powders prepared by high-energy ball milling and solid reaction products, X-ray diffraction (XRD) analysis was performed using a 'Difrey 401' diffractometer ('Scientific Instruments' JSC, Saint Petersburg, Russia) with Cr-Kα radiation (wavelength of 0.22909 nm). The XRD patterns were processed using a Powder Diffraction File™ database (PDF®) from the International Centre for Diffraction Data (ICDD).

Visual imaging investigation and particle size measurements of the manufactured powder samples were performed using a Bio 6 model optical microscope equipped with a high-resolution camera (UCMOS 10000KPA; 'Altami' LLC, Saint Petersburg, Russia). Particle size measurements in the microscope images were carried out by means of 'Altami Studio 3.5' software. The image processing procedure included 'capturing' particles in the image by adjusting rendering settings, contouring, and further calculation of the contour sizes (maximum Feret diameters) using the preliminary obtained calibration data.

The surface morphology of the manufactured powder samples and reaction products was investigated by scanning electron microscopy (SEM) in backscattered electron (BSE) imaging mode. For these investigations, a NOVA NanoSem 650 scanning electron microscope (FEI Co., Hillsboro, OR, USA) with an annular backscattered electron detector was used.

3. Results and Discussion

3.1. Characteristics of Hydroreactive Powders

3.1.1. X-ray Diffraction Analysis

For the experiments, powder samples of four different types were prepared by ball milling. A first sample contained metal scrap (80 wt.%) and KCl salt (20 wt.%); a second sample was composed of metal scrap (80 wt.%), KCl, and Wood's metal alloy (10 wt.%); a third sample was made of metal scrap (80 wt.%) and Wood's metal alloy (20 wt.%); and a fourth sample was metal scrap (100 wt.%) without additives.

The XRD patterns for the powder samples supplemented with ones for the original materials (metal scrap, salt, and alloy) are given in Figure 3. It was concluded from the XRD analysis results that all samples contained a solid solution of aluminum in the primary hexagonal close-packed magnesium lattice and face-centered cubic lattice of the aluminum phase. ML5 is a Mg-Al-Zn-type alloy with concentrations of the major elements generally close to those of the AZ91 alloy, which is composed mainly of α-phase Mg and β-phase $Mg_{17}Al_{12}$ [53]. Presumably, the identification of Al in the ML5 sample may be associated with the effect of the enrichment of the $Mg_{17}Al_{12}$ phase with Al after a thermal treatment (holding at 420 °C for 12 h and aging at 200 °C for 8 h) standard for this alloy grade. In report [54], the content of $Mg_{17}Al_{12}$ in an ML5 alloy sample was 5.31 wt.%, with the content of Al in the $Mg_{17}Al_{12}$ phase achieving as much as 3.42 wt.% of the total alloy sample mass.

For the samples ball-milled with salt, a KCl phase was observed. From a comparison of the XRD patterns for the powder samples with those for pure KCl, it can be seen that several KCl peaks were barely distinguishable in the XRD patterns of the powders.

Wood's metal alloy is composed of Pb_7Bi_3 solid solution (hexagonal close-packed lattice), Bi (rhombohedral lattice), Sn (tetragonal lattice), and Cd (hexagonal lattice) phases. The powders comprising Wood's metal alloy have an Mg_2Sn phase, which apparently originated from ball milling under increased temperature due to heat release. Thus, the presence of Sn from Wood's alloy in the powder samples was confirmed, while no other Wood's alloy components were detected by the XRD analysis.

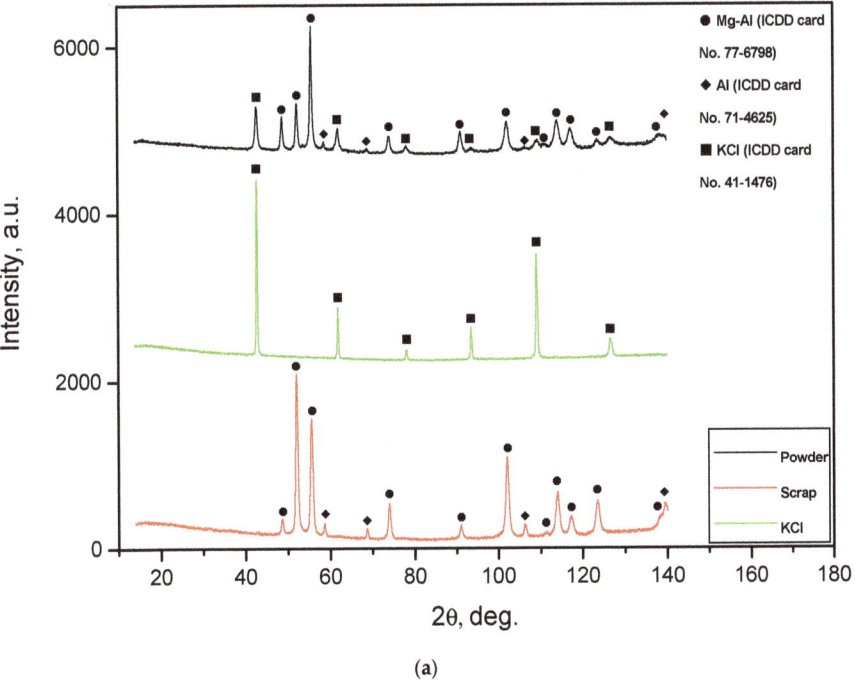

(a)

Figure 3. *Cont.*

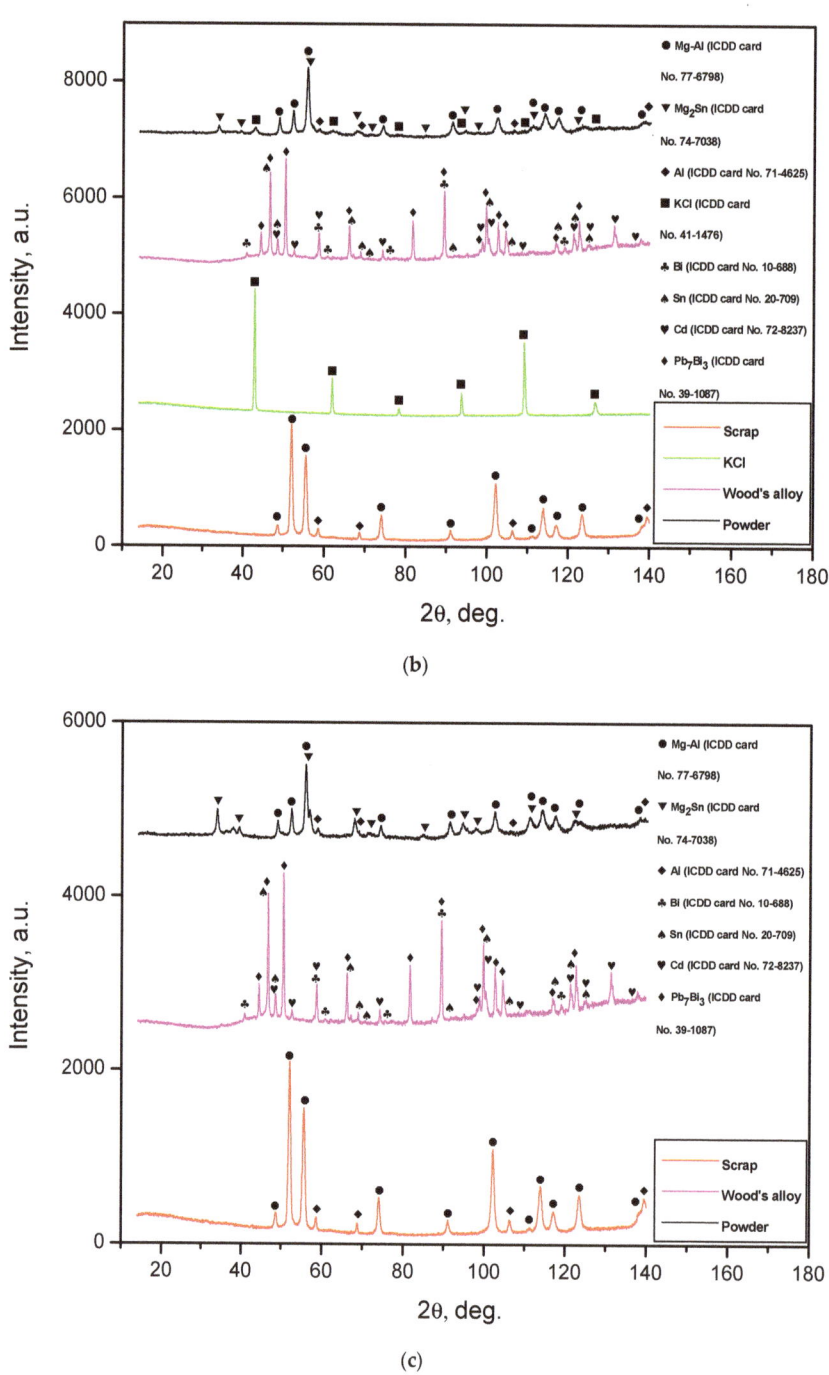

Figure 3. *Cont.*

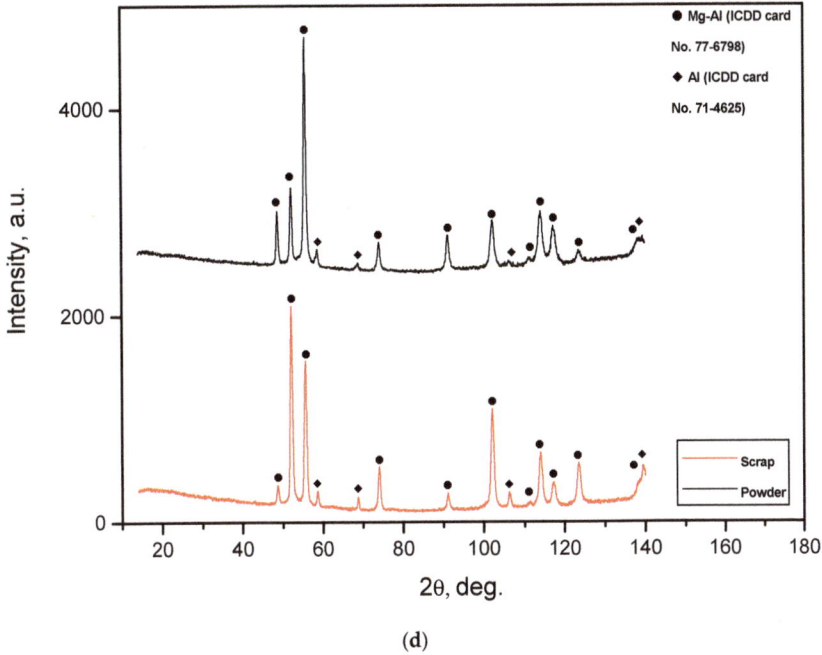

Figure 3. XRD patterns of the ball-milled powder samples: (**a**) Mg-Al (80 wt.%) and KCl (20 wt.%); (**b**) Mg-Al (80 wt.%), Wood's metal alloy (10 wt.%), and KCl (10 wt.%); (**c**) Mg-Al (80 wt.%) and Wood's metal alloy (20 wt.%); (**d**) Mg-Al (100 wt.%) without additives.

It can be seen that the peak intensities of the Mg-Al and Al phases for several metal scrap peaks were higher than those for the powder samples. Such a result can be ascribed to a partially textured structure of the original metal scrap (the orientation of crystalline grains along a certain crystallographic direction). A slight widening of the diffraction peaks for the Mg-Al phase was observed: the peaks for the sample of Al-Mg and Wood's metal alloy were wider than those for the sample of Mg-Al, KCl, and Wood's metal alloy, and for the latter, they were wider than those for both the Mg-Al and KCl sample and Mg-Al without additives. Such an effect obviously resulted from the difference in the crystallite sizes of the samples. It was demonstrated that, during ball milling, hard and brittle salt particles were fractured and contributed greatly to 'cutting' ductile Al-Mg particles into pieces with crystalline size reduction [48], while Wood's alloy did not produce such an impact. In a preceding study, it was demonstrated as well that, due to their higher hardness, KCl particles provided a greater enhancement of ball-milled Mg hydrogen production properties than NaCl [43] (which is why, in this study, KCl was used for ball milling).

For all the samples, no Fe contamination from ball milling with steel balls was detected by the XRD analysis. However, it was estimated from the change in the ball weight: the initial ball mass was 96.6050 g; after the first ball-milling cycle, it reduced to 96.5999 g; and after four cycles of ball milling, it decreased to 96.5863 g. Therefore, the average ball mass reduction per ball-milling cycle was about 0.005 g per 4 g of powder sample, which corresponded to the Fe content of ~0.1 wt.%. The estimated value is in good agreement with the iron contamination result (up to 0.1 wt.%) for the ball-milled Mg powders measured in study [26].

3.1.2. Investigation by Optical Microscopy

The photographs of the powder samples and original metal scrap captured using a microscope camera are given in Figure 4. As it can be seen from the photographs under ×20 magnification, the particles for all the powder samples had uneven, 'hummocky' surfaces. Such morphology resulted from the impact of steel balls (and KCl particles for the samples containing salt). The ball-milled sample with 20 wt.% Wood's alloy represented a dense layer covering the milling pot internal surface that was separated by scratching. From this fact, it was concluded that Wood's alloy liquefied during ball milling and produced a negligible impacting effect, if any. In the images of the metal scrap, the particle surfaces covered with scratches and uneven edges—resulting from mechanical processing—are shown.

From the other photographs (×2 and ×5 magnifications), it can be seen that, in general, the finest particles corresponded to the sample ball-milled with 20 wt.% KCl. The largest particles were ones of the sample ball-milled with 20 wt.% Wood's metal alloy and the 100 wt.% Mg-Al powder, and the particles of the sample containing both KCl and Wood's alloy had intermediate sizes, as expected. The finest particles demonstrated a tendency to form clusters (due to electrostatic adhesion), while the largest ones were loose. Metal scrap contained many particles too large to be captured in full size.

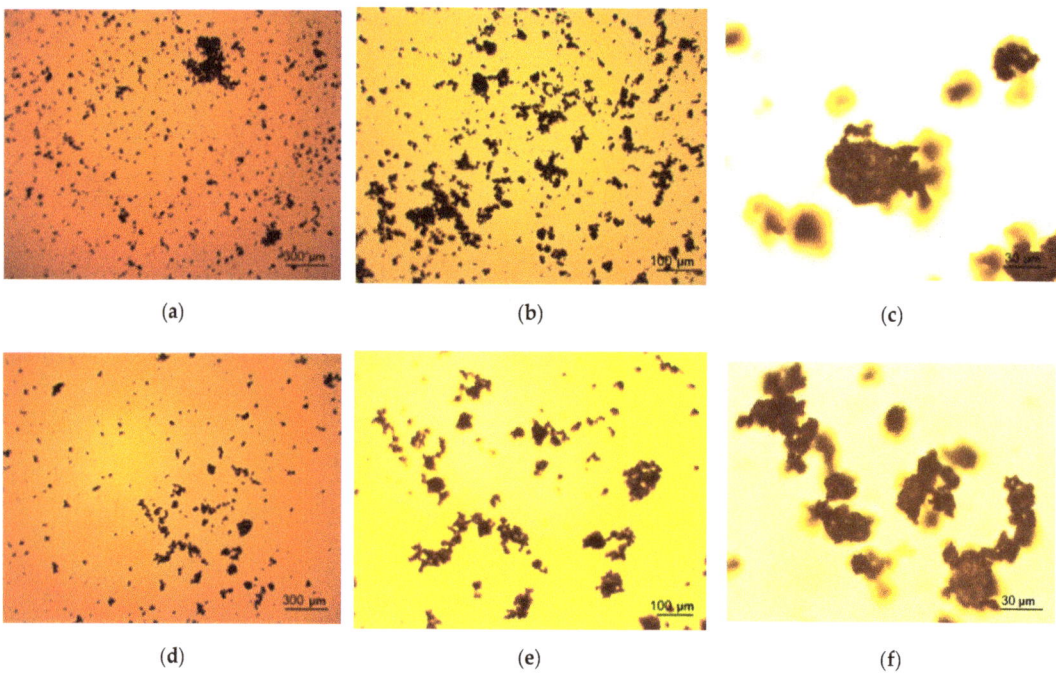

Figure 4. *Cont.*

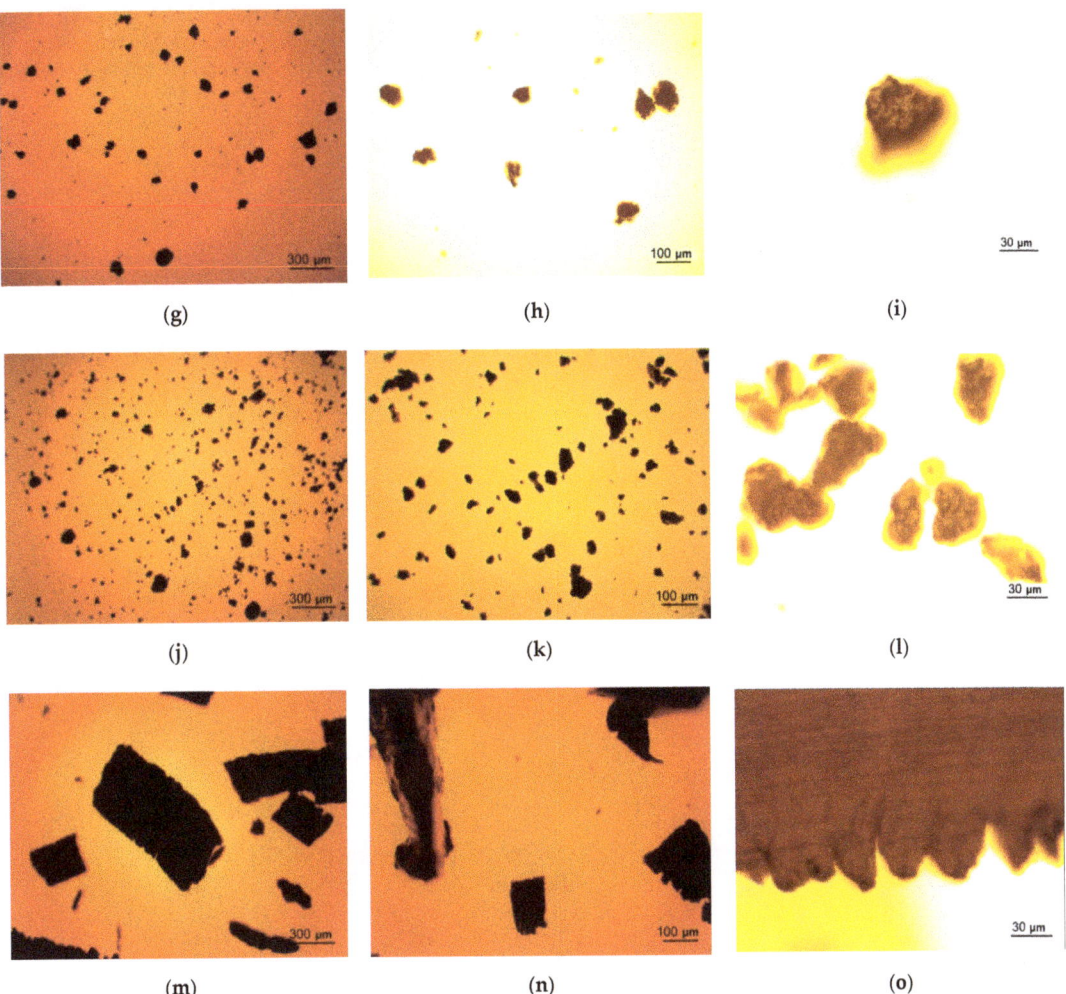

Figure 4. Microscope photographs of the ball-milled powder samples: (**a**–**c**) Mg-Al (80 wt.%) and KCl (20 wt.%) particles under ×2, ×5, and ×20 magnifications; (**d**–**f**) Mg-Al (80 wt.%), Wood's metal alloy (10 wt.%), and KCl (10 wt.%) particles under ×2, ×5, and x20 magnifications; (**g**–**i**) Mg-Al (80 wt.%) and Wood's metal alloy (20 wt.%) particles under ×2, ×5, and ×20 magnifications; (**j**–**l**) Mg-Al (100 wt.%) particles under ×2, ×5, and ×20 magnifications; (**m**–**o**) metal scrap particles under ×2, ×5, and ×20 magnifications.

Particle size distributions were obtained by processing several images of different random regions of the disperse samples. The numbers of images were 5 for each of the samples containing Wood's alloy, 10 for each of the samples both with KCl and without additives, and 20 for metal scrap. The corresponding histograms and cumulative curves are given in Figure 5. According to the obtained results, most of the particles for the sample containing 20 wt.% KCl had their sizes below 10 μm. Those for the sample with both KCl and Wood's alloy were mostly smaller than 20 μm. Particles of the powder with 20 wt.% Wood's alloy were generally limited to 35–40 μm. As to the sample obtained by scrap ball milling without additives (100 wt.% Mg-Al), its particle sizes were mostly smaller than

70–80 µm. The scrap sample was characterized by high non-uniformity in the particle size distribution with nearly 60% of the particles smaller than ~1 mm and the rest as large as ~1–10 mm. Such big sizes were attributed to the prolate form typical for a large portion of scrap particles.

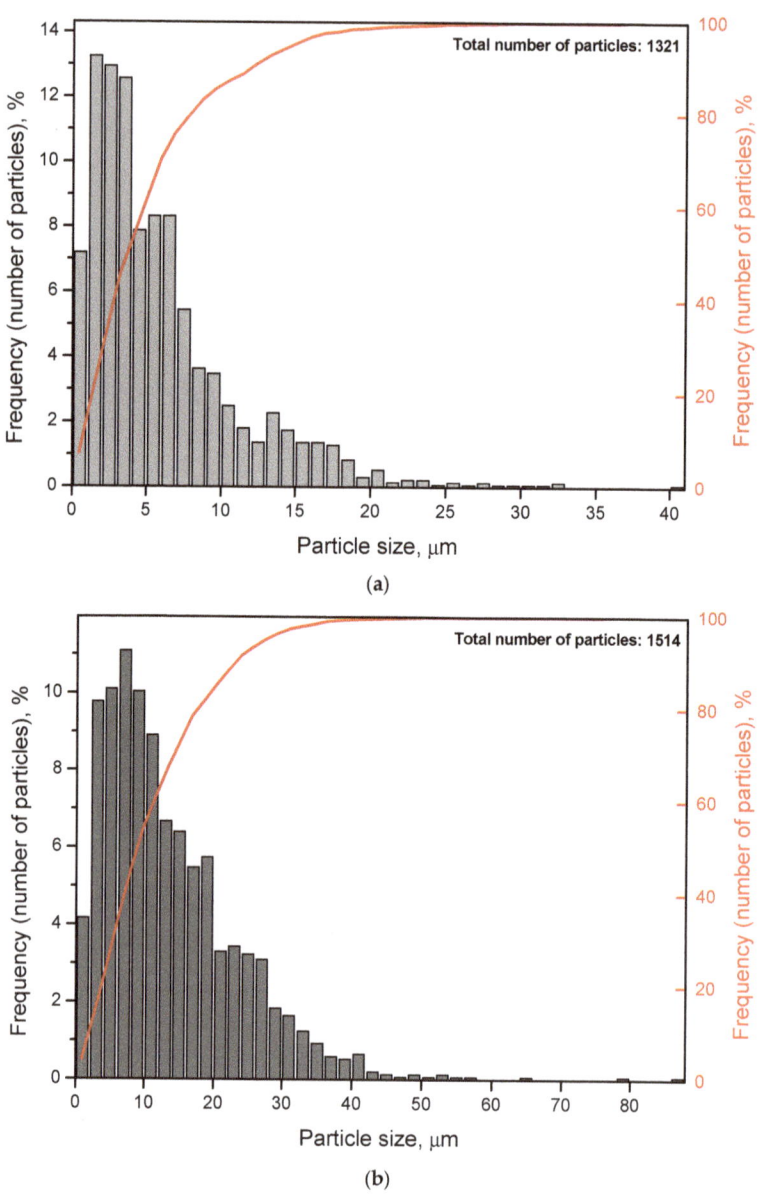

Figure 5. *Cont.*

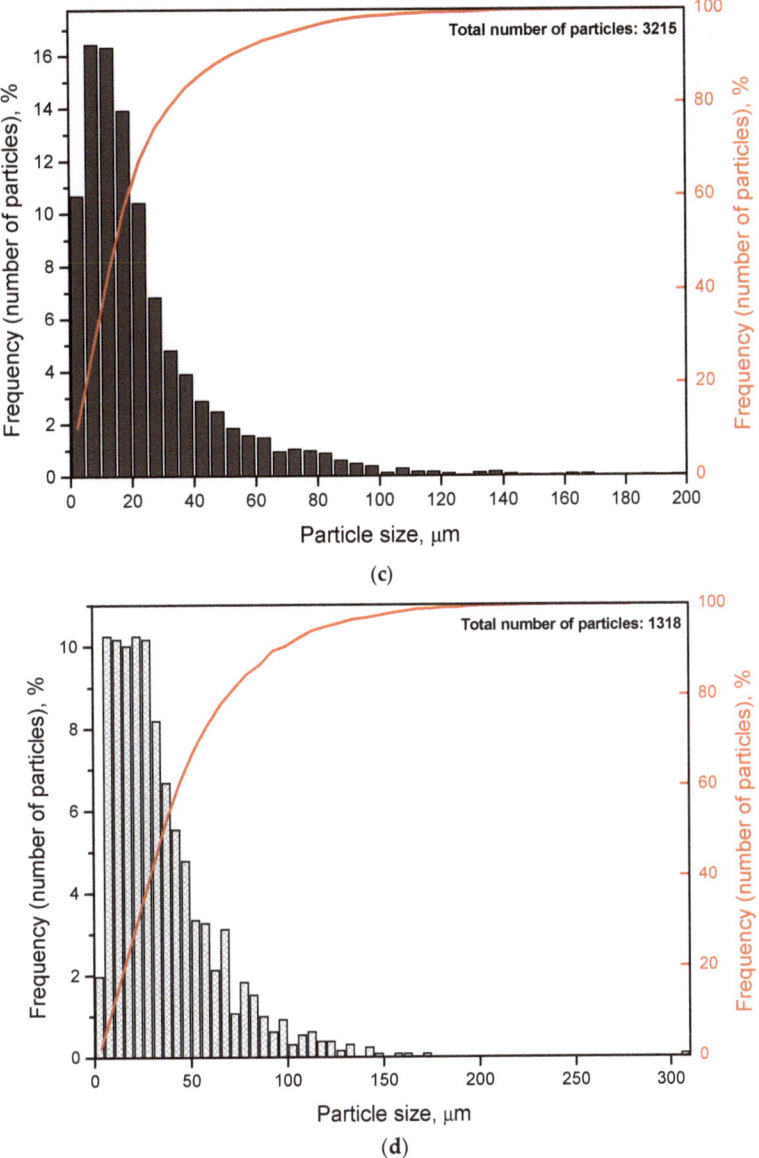

Figure 5. Cont.

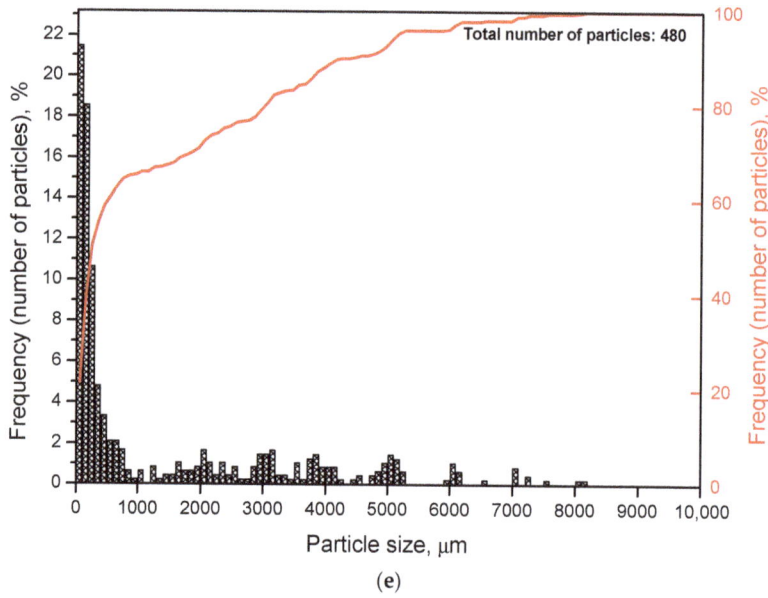

Figure 5. Particle size distributions (histograms and cumulative curves) for different samples: (**a**) Mg-Al (80 wt.%) and KCl (20 wt.%); (**b**) Mg-Al (80 wt.%), Wood's metal alloy (10 wt.%), and KCl (10 wt.%); (**c**) Mg-Al (80 wt.%) and Wood's metal alloy (20 wt.%); (**d**) Mg-Al (100 wt.%) without additives; (**e**) metal scrap.

The average particle sizes for the samples containing KCl, both Wood's alloy and KCl, Wood's alloy, and no additives were 6 µm, 14 µm, 24 µm, and 36 µm, respectively. Some errors in size determination could arise from the fact that the samples containing KCl might contain some amount of separate salt particles of the smallest sizes contributing to composite particle size 'underestimation'. The average scrap particle size was ~1.4 mm. The total numbers of particles used for size averaging are given in Figure 5.

According to the results of the optical microscopy investigations, the sample containing 20 wt.% KCl represented the finest powder and, consequently, it should have the largest specific surface area. The composition containing both KCl and Wood's alloy with larger particles should have a lower value of specific surface area. The powder containing 20 wt.% Wood's alloy was characterized by relatively big particles and should have a still smaller specific surface area. Finally, the sample of metal scrap ball-milled without additives comprised the largest particles and should have the smallest specific surface area.

3.1.3. Investigation by Scanning Electron Microscopy

The microphotographs of the powder samples obtained by scanning electron microscopy in BSE imaging mode are given in Figure 6. In the image of a particle composed of metal scrap (shown in grey and dark grey) and KCl (shown in light grey and white), embedded salt crystals and many cavities originating from impacts with salt particles were clearly seen on the surface. Particles composed of metal scrap (shown in grey and dark grey), KCl (shown in grey and light grey), and Wood's alloy (shown in white) also had some cavities on their surfaces, as well as embedded salt crystals and alloy spots. A particle of metal scrap (shown in grey and dark grey) with Wood's alloy (shown in light grey and white) likely represented a lamellar structure composed of smaller flattened particles agglomerated due to the 'cold welding' process during ball milling. Particles of the scrap ball-milled without additives represented similar multilayer structures formed by 'cold welding', with the magnesium-aluminum alloy shown in grey and inclusions of heavier elements—Zn, Zr, and Nd segregated on the surface and Fe contaminations from the steel balls—shown in light grey and white.

Figure 6. Cont.

(c)

(d)

Figure 6. BSE images of ball-milled powder samples: (**a**) Mg-Al (80 wt.%) and KCl (20 wt.%); (**b**) Mg-Al (80 wt.%), Wood's alloy (10 wt.%), and KCl (10 wt.%); (**c**) Mg-Al (80 wt.%) and Wood's alloy (20 wt.%); (**d**) Mg-Al (100 wt.%) without additives.

3.2. Hydrogen Evolution Kinetics

3.2.1. Metal Scrap and KCl

The preliminary tests on the oxidation of the powder samples in a 3.5 wt.% NaCl aqueous solution demonstrated that the sample composed of Mg-Al (80 wt.%) and KCl

(20 wt.%) started to react with vapor at nearly 80 °C. The sample inflamed and—a few seconds after emergency unsealing of the reactor loading outlet—a flame jet representing, presumably, burning magnesium and a hydrogen-air mixture was observed during several seconds. At 50 °C, the same composition inflamed upon contacting the surface of the aqueous solution. Additionally, a just-prepared sample with the same composition inflamed upon exposure to air. More careful pretesting of the samples (0.5 g weight portions) in the temperature range from 25 to 40 °C with an interval of 5 °C revealed that, for the samples containing KCl, ignition with a flash was observed at 40 °C. Therefore, hydrogen evolution kinetics was studied in the temperature range from 15 to 35 °C; before the experiments, the 'fresh' samples with 20 wt.% KCl were held in a glove box under Ar.

The kinetic curves obtained for the powder of Mg-Al and KCl are represented in Figure 7. As it can be seen, in the beginning of the reaction, the hydrogen evolution rate at 35 °C was the highest, as expected, and it decreased with lowering temperature. At all temperatures, the highest reaction rates were observed for nearly the first 20 min. of the experiment. At about 40 min. after the beginning, the kinetic curves tended to move towards their plateaus, reflecting a definite slowing down of the reaction.

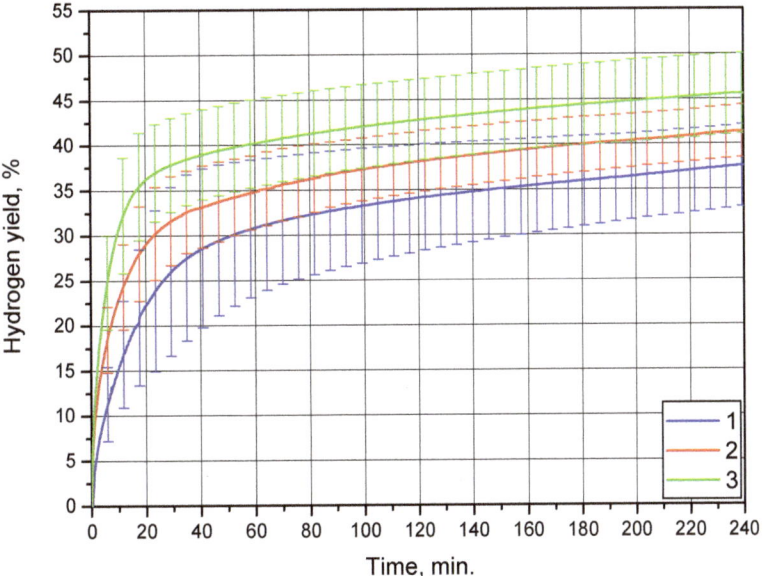

Figure 7. Hydrogen evolution kinetic curves for the Mg-Al (80 wt.%) and KCl (20 wt.%) powder: 1—15 °C, 2—25 °C, and 3—35 °C.

As for the hydrogen yields, large uncertainties were typical for all the kinetic curves. Moreover, overlapping of the error bars (standard deviations) of the kinetic curves was observed. After nearly 20 min. of the experiments, the error bars of all three kinetic curves overlapped within some yield intervals; by the end of the measurement time, the 'shared' yield interval was from 41.2% to 42.1%. The hydrogen yields obtained at 15, 25, and 35 °C were (37.6 ± 4.5)%, (41.4 ± 2.9)%, and (45.6 ± 4.4)%, respectively.

Such large variations in the experimental data probably resulted from the fast aging of this sort of powder materials, despite their storing under an Ar atmosphere. For a set of three experiments at different temperatures, the same powder sample was divided into portions, and the reactivity of the samples that were used later could decrease after several hours of storing.

In study [47], it was reported that the effect of salt particles on the activation of aluminum powders prepared by ball milling was that they acted as nano-millers. The aluminum activation mechanism constituted the creation of a great number of new 'fresh' metal surfaces and the prevention of their re-oxidation in air by the formation of salt layers on them. However, in study [45], it was demonstrated that the mechanism of merely covering aluminum particles with salt did not explain the higher reactivity of the samples produced by prolonged (4–19 h) ball milling. The said samples represented composite particles formed by the aggregation of the salt and aluminum phases. In hot water, salt inclusions dissolved, leaving many voids and tunnels in the aluminum particles. Such porous structure provided a good deal of 'fresh surfaces' coming into contact with hot water. Other effects of 'salt-assisted' ball milling useful for enhancing aluminum (or magnesium) corrosion with hydrogen evolution include an increase in lattice strain, a reduction in crystalline size, and the creation of defects (dislocations, vacancies, grain boundaries, etc.) in metal particles [48].

In the present study, Mg-based alloys were employed, and it seemed to be useful to compare the obtained results with similar ones for Mg and KCl composites. In study [43], samples manufactured with Mg powder (99.8% purity, particle size 20–100 mesh) and KCl were tested in pure water at 80 °C. The tested parameters closest to those employed in our study were a KCl concentration of 25 wt.% and 3 or 7 h of ball milling. According to the reported kinetic curves, the hydrogen yields for those samples achieved nearly 48–49% (~450–460 mL) in 1 h. The most intensive hydrogen release proceeded during nearly the first 5–10 min. of the experiment, and after ~15–20 min., a significant slow-down was observed, with the kinetic curves gradually moving towards their plateaus. The hydrogen yields of ~42–46% obtained in the present study at 25–35 °C are not drastically lower than the 48–49% obtained in [43] for Mg at 80 °C, considering that, by the end of the experiments (4 h and 1 h), the reaction either achieved its termination stage or scarcely proceeded. Somewhat higher hydrogen yields reported in [43] could be ascribed to a higher KCl content (25 wt.% vs. 20 wt.% in the present study), smaller original Mg powder particle size (~100 μm vs. ~100–1000 μm in the present study), and higher Mg content in the Mg powder compared to the Mg-based scrap.

In order to follow up on the discussion, some other results were referred to and compared. The relevant experimental conditions and results of preceding studies are given in Table 1.

From a comparison of the above data, the following conclusion can be made. In the cases of ball-milled Mg or Al mixtures with some salts (NaCl and KCl), the eventual hydrogen yields were majorly affected by the salt content and ball-milling time. However, in a certain temperature range, the said values were scarcely affected or almost unaffected by the reaction temperature. Salt content and ball-milling time most notably affected powder particle size and structure (embedded salt inclusions) and, therefore, specific surface area of the samples. As it can be seen, the particle sizes reported in different studies, as well as in the present one, were much different, which can be attributed to different conditions of their measurement (optical or scanning electron microscopes, number of particles for averaging, inclusion or exclusion of separate salt particles, etc.). However, for the results obtained in the present study, the following deduction still can be made. Upon dissolution of KCl inclusions in the samples, the uncovered 'fresh' surfaces were likely to be covered with a mostly dense and 'compact' layer of poorly soluble reaction product (according to the XRD data the product was $Mg(OH)_2$). As the powder particle sizes were mostly too large compared to the thickness of the $Mg(OH)_2$ layer at the moment of a considerable reaction slowing down or termination, a significant amount of the Mg-Al alloy remained unreacted. It is quite possible that—in a certain temperature range—such 'terminal' thicknesses of the reaction product layers (for Al or Mg) were about the same, regardless particle sizes. Therefore, the eventual hydrogen yield values depended on the ratio of the 'terminal' product layer thickness to the initial particle size, or the wideness of the metal regions between hollows remaining in the particles after dissolution of the embedded salt inclusions. In other words, the eventual hydrogen yields were substantially limited by the said ratio, while the reaction temperature (in a certain interval) affected only whether the said values were achieved sooner or later.

Table 1. Experimental conditions and results of preceding studies.

Reactive Material	Ball-Milling Time, h	Powder Particle Size, μm	Aqueous Media	Temperature, °C	Reaction Time, min.	H_2 Yield,%	Source
Al (~75 μm), NaCl (NaCl to Al mole ratio 2–7)	15	~100 nm	Distilled water	55–90	~20	100	[47]
Al (~75 μm), NaCl (NaCl to Al mole ratio 1)				55–70		<65	
Al (~100 μm), NaCl (NaCl to Al mole ratio 0.1–1.5)	20	~50 nm	Distilled water	70	~40	70–90 (mole ratios 0.5–1)	[48]
Al (227 μm), NaCl (50 wt.%)		21–22 μm			~60 or less	~100	
Al (227 μm), NaCl (25 wt.%)	7–19	-	Distilled water	80		~50 or less	[45]
Al (227 μm), NaCl (75 wt.%)		-			~20 or less	100	
Al (227 μm), KCl (50 wt.%)		16–17 μm					
Mg (150–750 μm), NaCl (50 wt.%)	3 7 15	277 μm 65 μm 46 μm				~16 ~55 ~62	
Mg (150–750 μm), KCl (25 wt.%)	3 7 15	-	Distilled water	80	~60 or less	~48 ~49 ~60	[43]
Mg (150–750 μm), KCl (50 wt.%)	15	-				>90	
Mg (150–750 μm), KCl (75 wt.%)	15	-				100	

These conclusions are supported by the results of studies [55,56], wherein experiments with aluminum nanoparticles (87 and 120 nm) provided their 100% conversion into hydrogen in pure water at 67–70 °C. In [56], the term 'penetration thickness' was introduced, and its values were calculated for various particle sizes and temperatures. According to the reported results, the penetration thicknesses for aluminum particles at temperatures up to ~100 °C were below 1 μm and not too much different from each other (taking into account the uncertainties). However, with increasing the temperature from 100 °C to 200 °C, these values rose up to ~2 μm. It was reported as well, that the penetration thickness was the same for different particle sizes, and its increase at temperatures over 100 °C was attributed to a considerable increase in diffusivity of the water. Therefore, the mentioned 'certain temperature range' is presumably limited by 100 °C or so.

3.2.2. Metal Scrap and Wood's Alloy

The results of the experiments with the Mg-Al (80 wt.%) and Wood's alloy (20 wt.%) composition are illustrated in Figure 8. The kinetic curves for this powder sample had an S-shape typical for topochemical reactions [52]. Kinetic curves for the said reactions usually have an 'acceleration' section in the beginning of the reaction, a section of the highest reaction rate with further deceleration, and a section referring to the completion of the reaction. The presence of a distinguishable 'acceleration stage' was accounted for slower initial reaction rates in the experiments for this powder compared with the previously discussed sample composition.

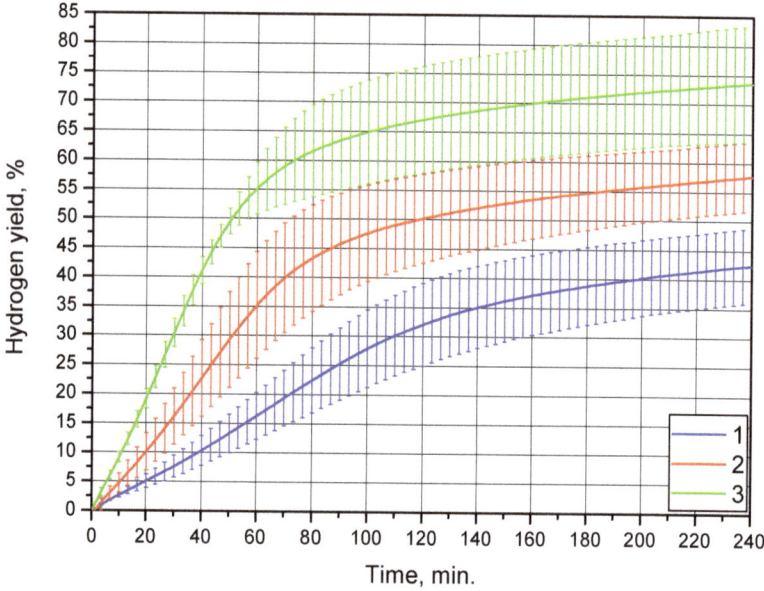

Figure 8. Hydrogen evolution kinetic curves for the Mg-Al (80 wt.%) and Wood's alloy (20 wt.%) powder: 1—15 °C, 2—25 °C, and 3—35 °C.

As in the preceding experiments, an increase in the hydrogen yield was observed, along with temperature growth. However, the differences between the hydrogen yield values for different temperatures were, in general, considerably larger than those for the previously discussed sample with salt. In study [57], some of the hydrogen yield statistical uncertainties for aluminum powder samples reacting with water achieved nearly 10%. Thus, relative deviation values of 10% or so may be typical for similar experiments with aluminum or magnesium powders. In the case of the present study, such statistical uncertainty may result from the variation in powder particle sizes, irregular distribution of salt or alloy additives, structure defects produced during ball milling, and, presumably, the effect of powder aging.

Although the uncertainties were large for all the kinetic curves, no overlapping of their error bars was observed, except at the very beginning of the reaction. Therefore, the results for this sample were quite distinguishable. The hydrogen yields achieved were (42.4 ± 6.3)%, (57.7 ± 5.9)%, and (73.5 ± 10.0)% for 15, 25, and 35 °C, respectively. These values were substantially higher than those for the sample with 20 wt.% KCl salt.

The addition of Bi and Sn to aluminum and magnesium is rather a common method to promote their reactions with water [5,37,58,59]. The standard electrode potentials of all Wood's alloy components—Bi, Sn, Cd, and (presumably) Pb_7Bi_3—and the intermetallide-phase $MgSn_2$ detected in the ball-milled powder samples were higher than those of Mg and Al (although some shifts in the 3.5 wt.% NaCl solution can be expected). Therefore, in the conductive media, the dominating mechanism of the hydrogen producing reaction was 'anodic dissolution' of the Mg (and probably Al) due to galvanic corrosion. Although an electrode potential difference between Mg and Al (or its intermetallide $Mg_{12}Al_{17}$) also promotes oxidation of the less noble metal (Mg) in conductive (0.6 mol/l NaCl solution) media [60], the said potential difference is much less than that between Mg and the Wood's metal components.

3.2.3. Metal Scrap, KCl, and Wood's Alloy

The results of another set of experiments for the powder of Mg-Al (80 wt.%), Wood's alloy (10 wt.%), and KCl (10 wt.%) are given in Figure 9. The represented kinetic curves demonstrated a predictable increase in the reaction rate along with the temperature growth. As compared to the preceding results, 'acceleration' sections of the kinetic curves were not visible, which indicated a relatively fast start of the reaction. The error bars for all three kinetic curves were much shorter than those for the previously discussed samples. This suggests a surprisingly high reproducibility of the experimental data for this composite material. It can be supposed that either such results were obtained by coincidence and more experiments are required for data averaging or that the material had some specific properties whose investigation is beyond the scope of the present research.

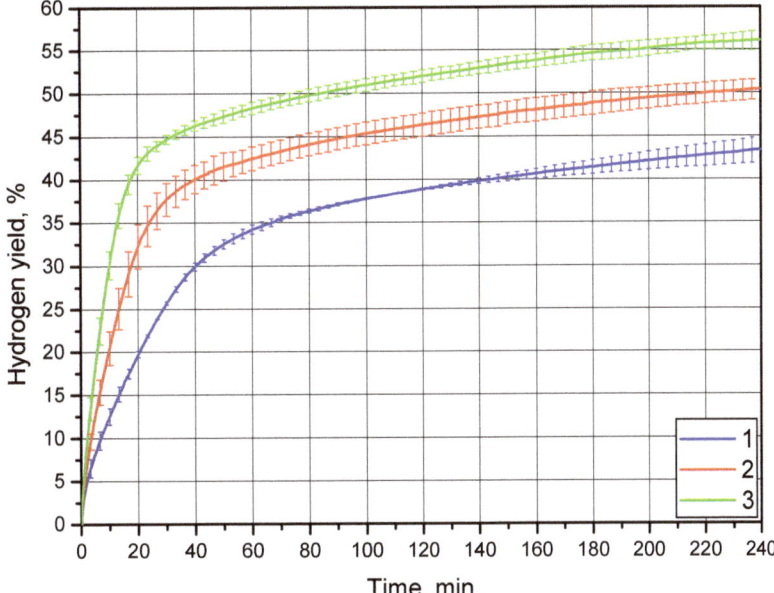

Figure 9. Hydrogen evolution kinetic curves for the Mg-Al (80 wt.%), Wood's alloy (10 wt.%), and KCl (10 wt.%) powder: 1—15 °C, 2—25 °C, and 3—35 °C.

For this powder sample, a synergetic effect of galvanic corrosion from Wood's alloy and specific surface area enlargement from KCl was expected to provide the highest hydrogen yields and evolution rates compared to the samples with either of these additives. The time intervals until the reaction deceleration for this sample were nearly 20–40 min. at 25 and 35 °C and 40–60 min. at 15 °C. The said time periods were longer than those for the sample with 20 wt.% KCl—up to 20 min at 35 °C and 20–40 min. at 15 and 25 °C. Additionally, these intervals were shorter than those for the sample with 20 wt.% Wood's alloy—nearly 60–100 min. for 25 and 35 °C and 120–160 min. at 15 °C. The hydrogen yields achieved were $(43.3 \pm 1.5)\%$, $(50.4 \pm 1.2)\%$, and $(56.0 \pm 1.2)\%$ at 15, 25, and 35 °C, respectively. These results exceeded those for the sample with the KCl additive and fell below ones for the Wood's alloy additive. Therefore, the results for hydrogen yields and initial reaction rates for the sample with both additives appeared to fall between those for the samples containing either of the two additives.

3.2.4. Metal Scrap Powder without Additives

Figure 10 demonstrates the results obtained for the sample of metal scrap ball milled without additives. Although some error bars for the kinetic curves corresponding to 25 and 35 °C partially overlapped, all the curves were still easily distinguishable. An S-shape was clearly identified for the curve at 15 °C, while for the other curves, it was too faint. The kinetic curve sections of reaction deceleration were located in the intervals of 20–60 min., 40–80 min., and 60–100 min. for 35, 25, and 15 °C, respectively. The hydrogen yields corresponding to 15, 25, and 35 °C were (56.0 ± 1.1)%, (64.0 ± 5.1)%, and (70.6 ± 2.5)%, respectively. Without any doubts, these values were much higher than those for the sample ball-milled with 20 wt.% KCl. Additionally, they definitely exceeded the corresponding hydrogen yield figures for the sample comprising both KCl and Wood's alloy. As to a comparison with the sample containing 20 wt.% Wood's alloy, at 15 °C the hydrogen yield for the latter was lower, while at 25 and 35 °C, these values matched within the statistical error limits.

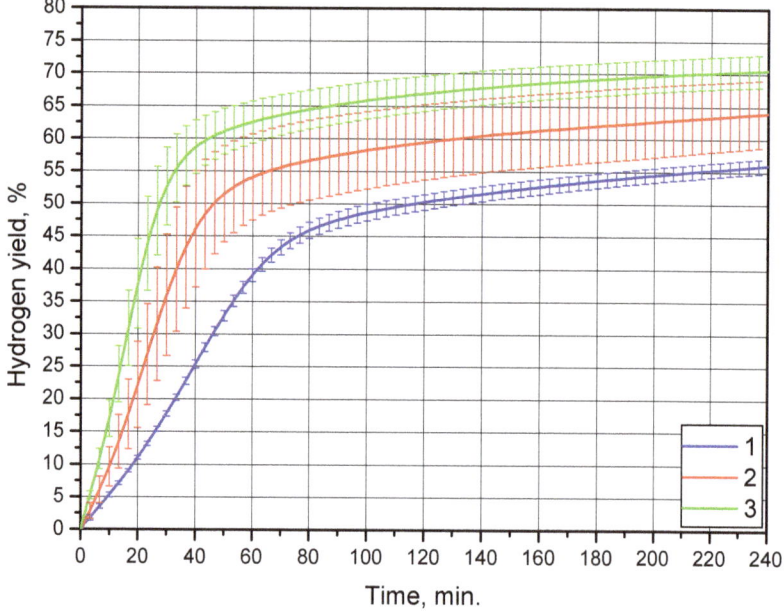

Figure 10. Hydrogen evolution kinetic curves for the Mg-Al (100 wt.%) powder: 1—15 °C, 2—25 °C, and 3—35 °C.

For this sample, the mechanisms presumably enhancing Mg oxidation in the NaCl aqueous solution were based on the formation of imperfections (dislocations, vacancies, grain boundaries, stacking faults, etc.) in the metal lattice due to ball milling [48]. Such structure defects favor the localized enrichment of Cl^- ions; therefore, they are readily attacked by pitting corrosion [25]. Another useful mechanism can be associated with the formation of porous lamellar structures with interlayer spaces at certain ball-milling durations. Such structures are composed of aggregated flattened particles, and despite large sizes of the entire aggregates, their specific surface area can achieve rather high values [46]. Unfortunately, in the present study, measurements of the specific surface areas were not performed. SEM investigation of the cross-sections of powder particles for lamellar structures identification was out of the scope of this study as well.

However, some information on the accumulated strain energy was derived from comparison of the peaks of XRD patterns for different powder samples. In [48], it was reported that the Al powder milled with NaCl had more lattice strain compared to that milled without NaCl, which was proved by a comparison of the full width at half maximum for the typical Al phase peaks. A similar visual analysis of the peaks of the powder samples with different additives (and with no additives) was performed in the present study. In Figure 11, several typical peaks for the base Mg-Al phase in the XRD patterns for different powders are shown. According to the assessment results, the full width at half maximum for two less intensive peaks had nearly the same value for all powder samples. For a third—the most intensive—peak, the relevant width values for the samples with Wood's alloy and with no additives were close, while the width of the sample with 20 wt.% KCl was definitely smaller. Thus, the peak for the sample ball milled without salt turned out to be wider than the peak for the sample with 20 wt.% salt, and this contradicts the result from [48]. However, this contrariety could be attributed to a difference in the ball milling time (20 h vs. 4 h in the present study), materials (Al vs. Mg-based alloy in the present study), as well as implemented salts (NaCl vs. KCl in the present study) and their contents.

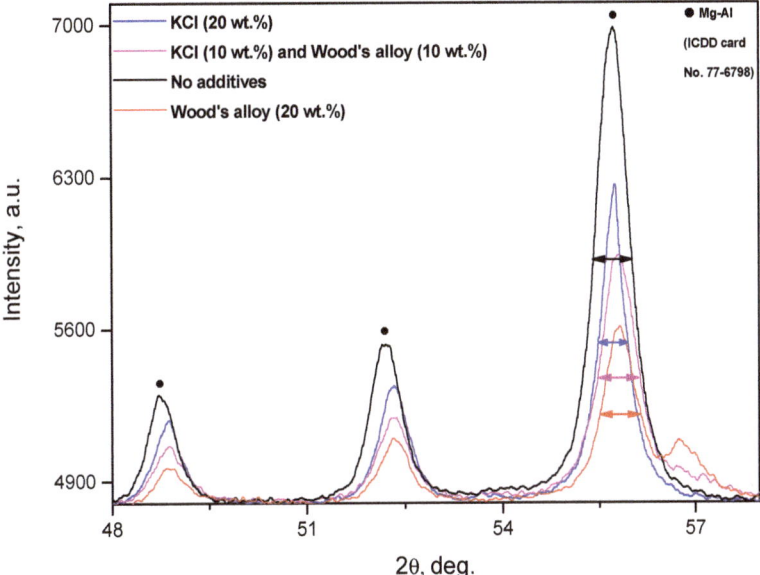

Figure 11. XRD patterns of ball-milled powder samples.

Therefore, it can be assumed that the latter sample with the narrowest peak width accumulated less strain energy in its crystal lattice during 4 h of ball milling compared to the other powder samples. Thus, for this sample, a large portion of ball-milling energy was actually employed for particle size reduction by cutting with salt particles, rather than for the creation of lattice imperfections. Consequently, this sample could be less affected by pitting corrosion from NaCl aqueous solution. Additionally, the effect of pitting corrosion could contribute significantly to sustaining the reaction.

The fact that the hydrogen yields for the powder without additives were close to or higher than those for the sample with 20 wt.% Wood's alloy can be explained from the following considerations. In some preceding studies, it has been demonstrated that some additives tended to form intermetallic compounds located at grain boundaries, screening the base metal in the grains from liquid. In cases of high concentrations, some intermetallic phases may hinder the reaction. For example, Cu is known to promote an Al–water reaction, however, its addition to Al-Ga-In-Sn systems inhibits the said reaction [61]. It was shown as well, that adding Ga to Mg promoted its corrosion in an NaCl solution [62], while adding

Ga to an Mg-Sn system hindered Mg corrosion in an NaCl solution [63]. Additionally, it was proved that for certain additives there are some optimal concentrations (for example, Sn in an Mg-Ni system [64]) corresponding to the highest hydrogen yields and evolution rates and that exceeding these optimal values led to a decrease in the said values. The abovementioned facts brought us to the conclusion that the contents of Wood's alloy in the samples tested in the present study were far from being optimal, and we suspect that the optimal values may actually be lower than the tested ones.

3.3. Characteristics of Solid Reaction Products

3.3.1. X-ray Diffraction Analysis

The XRD patterns for the solid reaction products obtained by the oxidation of different hydroreactive powder samples are given in Figure 12. The XRD analysis results demonstrated that all the reaction products contained some amount of the unreacted Mg-Al phase. The intensities of the corresponding peaks for the reaction products were lower than those for the powder samples before reaction, but most of the peaks were still distinguishable. In the XRD patterns for all the reaction product samples, $Mg(OH)_2$ was present. This phase resulted from the oxidation of Mg—the base component of the metal scrap—by an NaCl aqueous solution. A comparison of the XRD patterns for the powder samples with those for their reaction products showed a disappearance of the KCl phase. This was obviously attributed to KCl particle dissolution in the NaCl aqueous solution. In the samples comprising Wood's metal alloy, Mg_2Sn—present in the powders before the reaction—remained. $PbBiO_2Cl$ apparently originated from a reaction between the Wood's alloy component Pb_7Bi_3 and the NaCl aqueous solution. In a preceding study [28], for the experiments with 'mechanical alloy' powders—55 wt.% Mg, 30 wt.% Al, and 15 wt.% Fe ball-milled for 4 h—and 0.6 mol/l NaCl solution, in the solid reaction product $Mg_6Al_2(OH)_{18} \cdot 4.5H_2O$ and $Al(OH)_3$ were detected. However, from the present XRD patterns, neither Al nor its reaction products were identified. Such a result may be associated with a relatively low Al content in the ML5 alloy scrap—up to 9 wt.% according to the standard—or with the possible formation of pseudoboehmite, which has weak diffraction intensity.

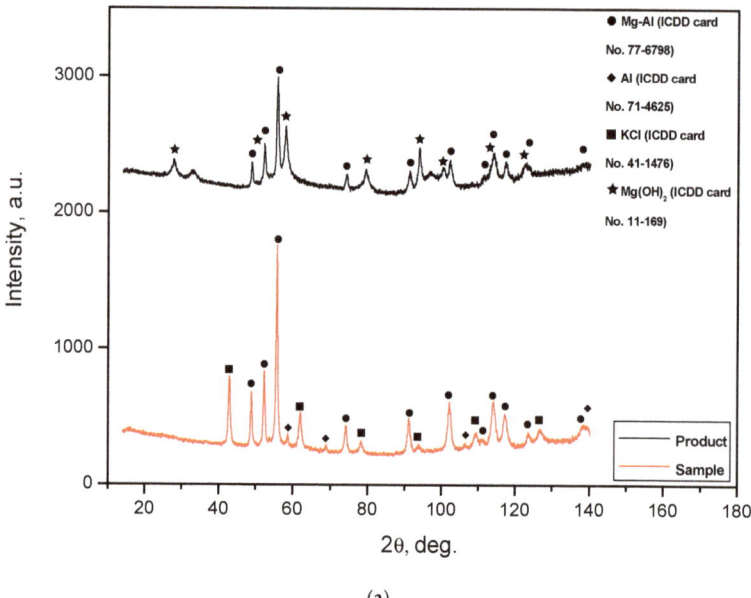

(a)

Figure 12. *Cont.*

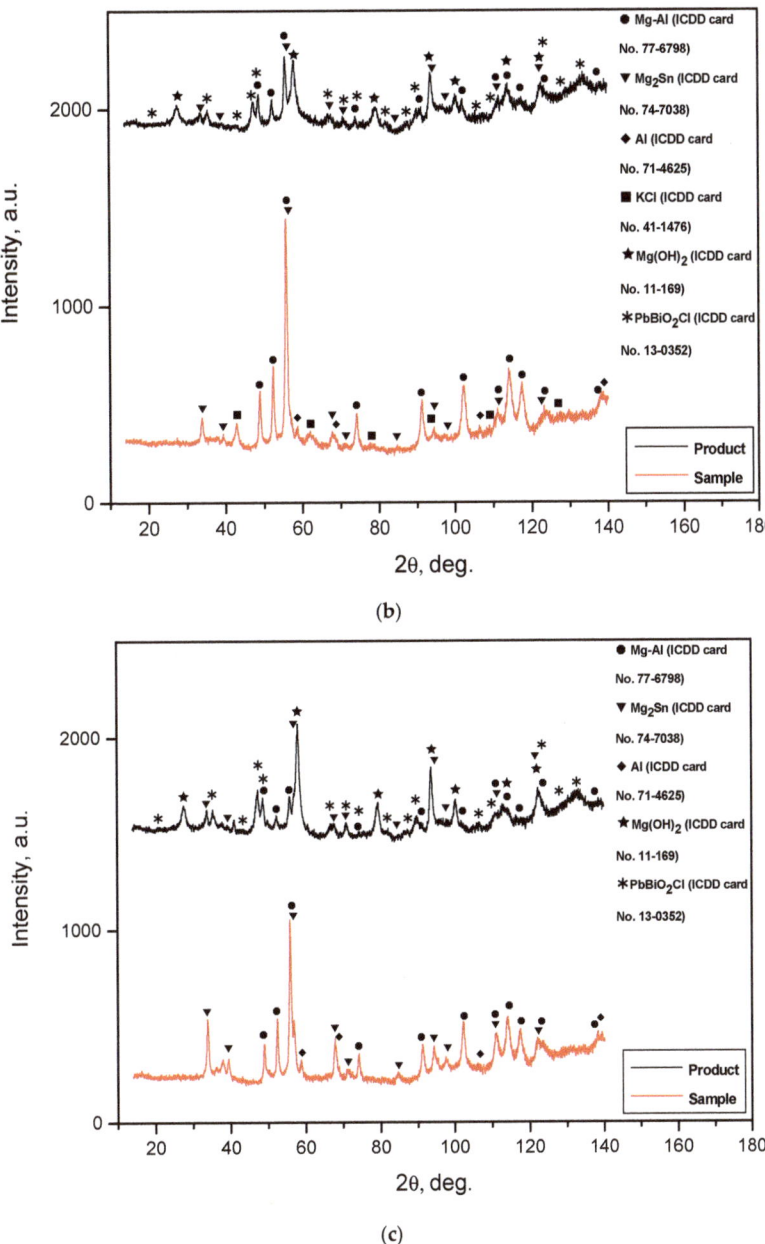

Figure 12. Cont.

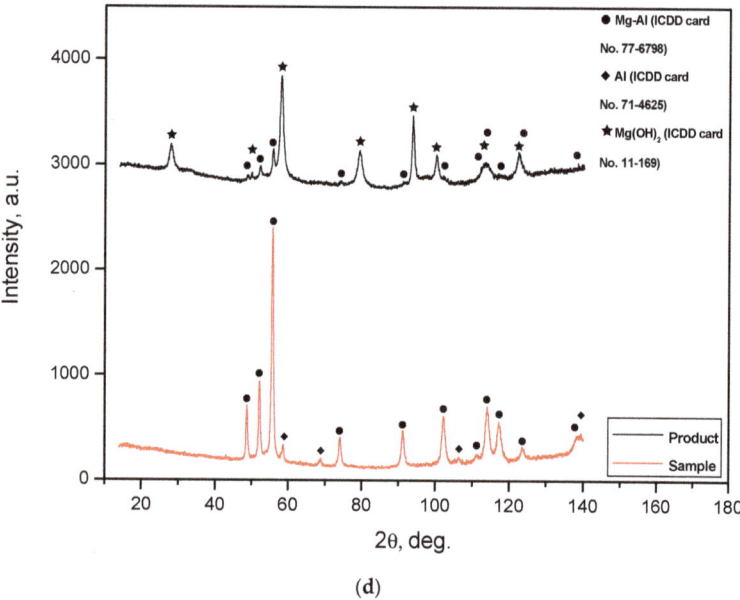

(d)

Figure 12. XRD patterns of the solid reaction products for different powder samples: (**a**) Mg-Al (80 wt.%) and KCl (20 wt.%); (**b**) Mg-Al (80 wt.%), Wood's metal alloy (10 wt.%), and KCl (10 wt.%); (**c**) Mg-Al (80 wt.%) and Wood's metal alloy (20 wt.%); (**d**) Mg-Al (100 wt.%) without additives.

3.3.2. Investigation by Scanning Electron Microscopy

The BSE images of the reaction product samples are represented in Figure 13. As it can be seen from the images, for the sample comprising metal scrap and KCl, the solid reaction product formed highly irregular structures on the particle surfaces. These irregular structures included clusters of large-sized 'flakes' (~100–500 nm), along with spongy areas formed by much smaller structure elements that looked similar to moss cover. The product layer corresponding to the sample manufactured with metal scrap, KCl, and Wood's alloy also had a very irregular morphology. The said morphology was characterized by porous structures formed of large-sized flakes and regions where the product layer formed in a much more 'compact' and dense manner. For the sample composed of metal scrap and Wood's alloy, the visible particle surfaces were totally covered with clusters composed of large-scale flakes. As for the sample manufactured of scrap without additives, not all of the captured particles were covered with large-sized flakes. The surfaces of the larger particles had quite irregular morphologies including 'mossy' areas, while almost all of the smaller particles looked like clusters of large-sized flakes.

It can be assumed that the structures composed of large-scale flakes were more liquid-permeable compared to the spongy areas formed by fine-structured elements. Although visual assessment of the sizes of large-scale flakes for different samples is not sufficient for making any final conclusions, the flakes for the samples containing Wood's alloy and for the sample without additives seemed to be generally larger than those for the sample with the KCl additive only.

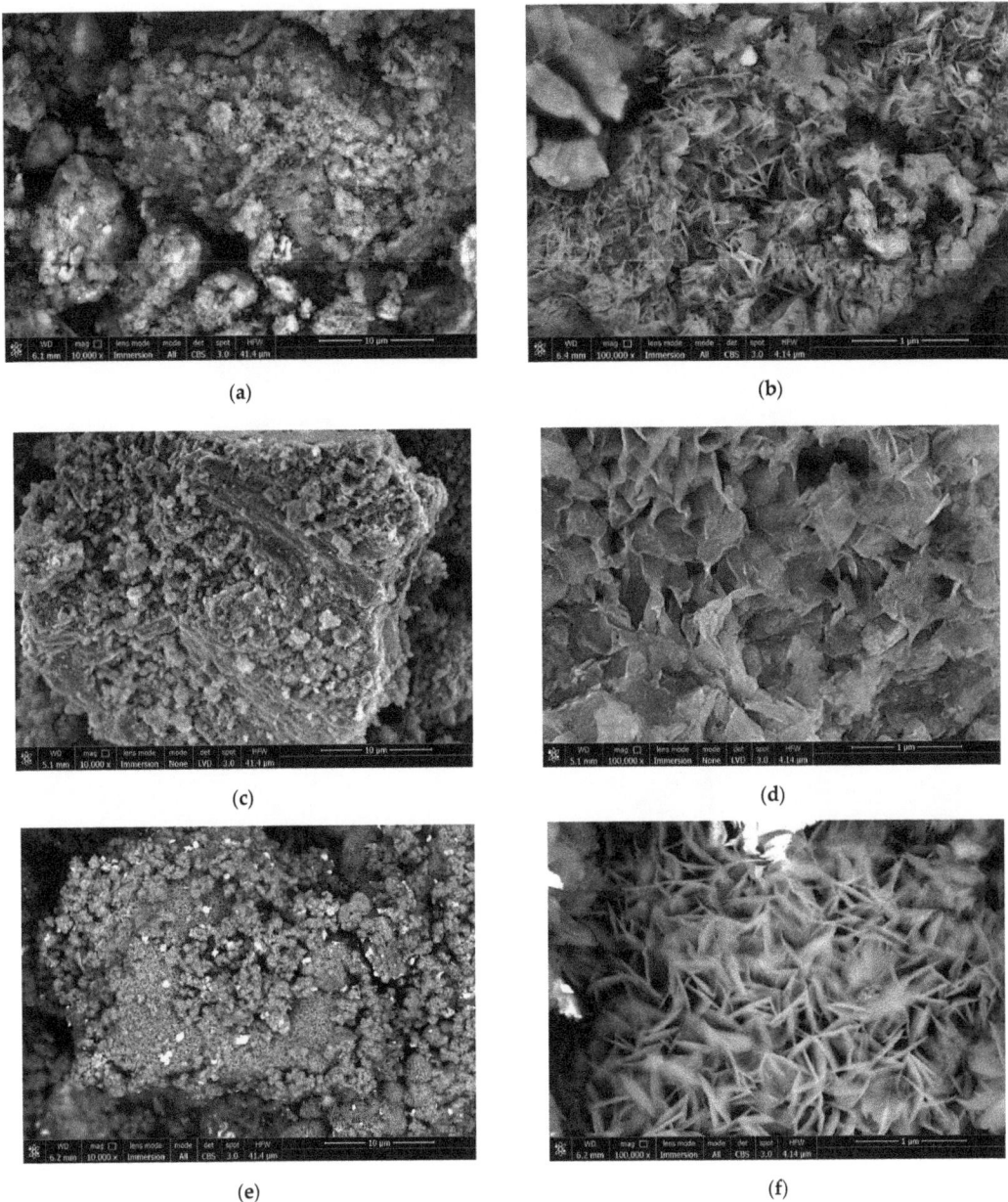

(a)

(b)

(c)

(d)

(e)

(f)

Figure 13. *Cont.*

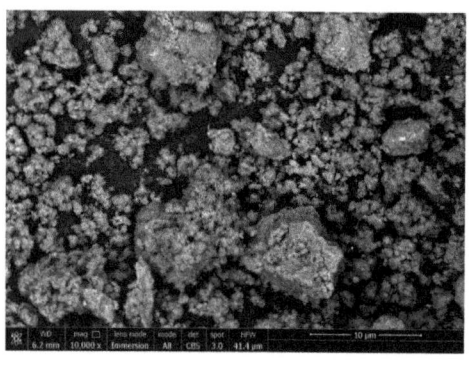

(g)   (h)

Figure 13. BSE images of the reaction products for different samples: (**a**,**b**) Mg-Al (80 wt.%) and KCl (20 wt.%); (**c**,**d**) Mg-Al (80 wt.%), Wood's alloy (10 wt.%), and KCl (10 wt.%); (**e**,**f**) Mg-Al (80 wt.%) and Wood's alloy (20 wt.%); (**g**,**h**) Mg-Al (100 wt.%) without additives.

3.4. Summarization and Discussion of the Results

The kinetic curves for all the experiments were approximated by Equation (2), and from those approximations, the values of the reaction rate constant (k) were derived. The obtained k values were used for graphical determination of the activation energy (E_a) from Arrhenius plots based on Equation (3). The Arrhenius plots for all powder samples are represented in Figure 14. The data on reaction constants and activation energy are given in Table 2. The calculated results demonstrated that the powder containing 20 wt.% KCl had the lowest activation energy of (7.0 ± 0.8) kJ/mol, the sample comprising 10 wt.% KCl and 10 wt.% Wood's alloy had a higher activation energy of (30.6 ± 2.9) kJ, the scrap ball-milled without additives had a somewhat higher activation energy of (35.9 ± 2.5) kJ, and the highest activation energy of (48.6 ± 4.8) kJ corresponded to the sample with 20 wt.% Wood's alloy.

Table 2. Activation energy and reaction constant values for different experiments.

Sample Composition	Reaction Rate Constant (k), s^{-1}	Reaction Parameter (n)	Activation Energy, kJ/mol
Mg-Al and KCl	2.4·10^{-4} (15 °C) 2.6·10^{-4} (25 °C) 2.9·10^{-4} (35 °C)	0.26 (15 °C) 0.21 (25 °C) 0.17 (35 °C)	7.0 ± 0.8
Mg-Al, Wood's metal and KCl	1.00·10^{-3} (15 °C) 1.42·10^{-3} (25 °C) 2.27·10^{-3} (35 °C)	0.35 (15 °C) 0.28 (25 °C) 0.24 (35 °C)	30.6 ± 2.9
Mg-Al and Wood's metal	2.33·10^{-3} (15 °C) 4.12·10^{-3} (25 °C) 8.73·10^{-3} (35 °C)	0.83 (15 °C) 0.66 (25 °C) 0.62 (35 °C)	48.6 ± 4.8
Mg-Al	3.85·10^{-3} (15 °C) 6.00·10^{-3} (25 °C) 10.02·10^{-3} (35 °C)	0.59 (15 °C) 0.46 (25 °C) 0.40 (35 °C)	35.9 ± 2.5

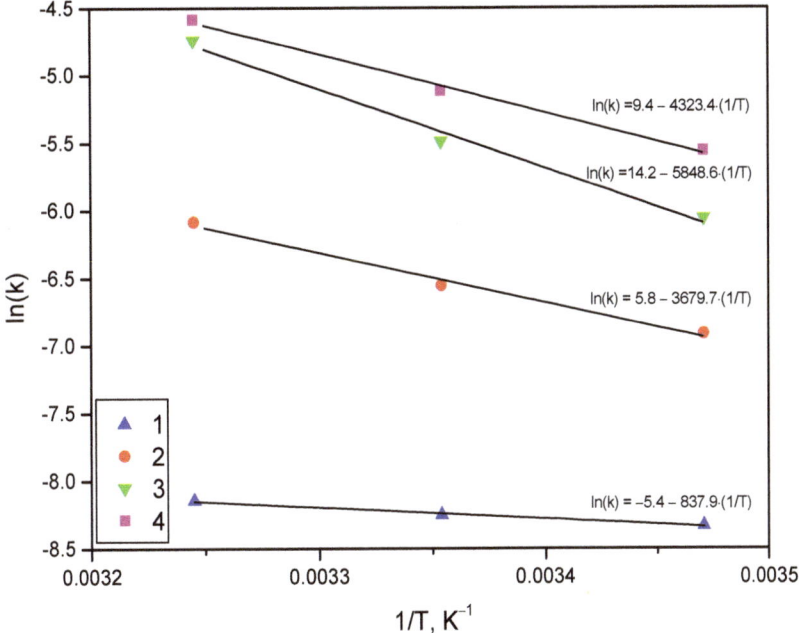

Figure 14. Arrhenius plots for different samples: 1—Mg-Al (80 wt.%) and KCl (20 wt.%); 2—Mg-Al (80 wt.%), Wood's alloy (10 wt.%), and KCl (10 wt.%); 3—Mg-Al (80 wt.%) and Wood's alloy (20 wt.%); 4—Mg-Al (100 wt.%) without additives.

For easier summarization of the study results and their discussion, the total hydrogen yields and maximum hydrogen evolution rates for all experiments were gathered in Table 3. The main findings based on the reaction kinetics data, results of sample analyses, and calculations were the following:

Table 3. Hydrogen production yields for different experiments.

Sample Composition	Tested Temperature, °C	Maximum H_2 Evolution Rate, mL/min./g	H_2 Yield,%
Mg-Al and KCl	15	36.1	37.6 ± 4.5
Mg-Al and KCl	25	69.1	41.4 ± 2.9
Mg-Al and KCl	35	83.0	45.6 ± 4.4
Mg-Al, Wood's metal, and KCl	15	28.0	43.3 ± 1.5
Mg-Al, Wood's metal, and KCl	25	38.1	50.4 ± 1.2
Mg-Al, Wood's metal, and KCl	35	50.0	56.0 ± 1.2
Mg-Al and Wood's metal	15	7.2	42.4 ± 6.3
Mg-Al and Wood's metal	25	9.8	57.7 ± 5.9
Mg-Al and Wood's metal	35	11.5	73.5 ± 10.0
Mg-Al	15	10.3	56.0 ± 1.1
Mg-Al	25	14.5	64.0 ± 5.1
Mg-Al	35	21.4	70.6 ± 2.5

- Powder containing 20 wt.% KCl provided the highest hydrogen evolution rate at the beginning of the reaction. It had the finest particles and, presumably, the largest specific surface area. Its activation energy was the lowest. The two latter facts were the reasons for the highest initial reaction rates. However, this sample provided the lowest hydrogen yield for 4 h of experimentation. Such an effect was attributed to

the formation of a 'dense' and 'compact' reaction product layer on the surface of the powder particles. The product layer structure included large, 'mossy' areas of fine-structured elements that were likely low-permeable to liquid with clusters of large-sized 'flakes', which, however, were not abundant at all;
- Powder activated with 20 wt.% Wood's alloy provided the lowest hydrogen evolution rates at the beginning. Its particle size was the second-largest, and the specific surface area was likely the third-largest among the four samples. It had the highest activation energy. The lowest initial reaction rate possibly resulted from the combination of a low specific surface area and high activation energy. However, the hydrogen yields for this sample were higher than those for the powders with KCl but comparable with those for the ball-milled scrap without additives. The high hydrogen yields were achieved due to the fact that the reaction product structure covering the particles was almost entirely formed by large-sized 'flakes'. Due to this, the product layer remained highly permeable to liquid for a long time during the reaction. It should be noted that the negligible effect of alloying for this sample (compared to the ball-milled scrap) could be attributed to possible screening effects from some of its compounds with Mg scrap due to extra concentration in the sample. The said effects could suppress the activation effect from the formation of microgalvanic couples of Mg and Wood's alloy components by ball milling;
- Powder with both 10 wt.% KCl and 10 wt.% Wood's alloy had its maximum hydrogen evolution rate exceeding that of the sample with 20 wt.% Wood's alloy and falling below that for the sample with 20 wt.% KCl. The particle size and hydrogen yield for this sample fell between these values for the powders with KCl and Wood's alloy. Its product layer morphology contained both 'mossy' areas typical for a sample activated with salt and large-sized 'flakes' presented on a particle surface of the reacted alloyed sample. Its activation energy was the second-lowest among the four samples;
- Powder of the scrap ball-milled without additives had the second-highest activation energy. Its maximum hydrogen evolution rate fell between the relevant values for the powder with 20 wt.% Wood's alloy and that with both (salt and alloy) additives. It had the largest particle size and, presumably, the smallest specific surface area. The hydrogen yields for this sample were close to or higher than those for the one with 20 wt.% Wood's alloy. Such an effect was attributed to a combination of the following factors: the accumulation of plenty of lattice imperfections favoring pitting corrosion from ball milling, the absence of any crystal grain screening compounds, and the formation of the product layer mostly in the form of clusters of large-sized 'flakes'.

Thus, under the considered reaction and ball-milling conditions, the implementation of KCl as an additive resulted in the lowest conversion of Mg-Al scrap into hydrogen. Moreover, this material required careful handling due to its combustible nature and demonstrated a high uncertainty for 'Hydrogen Yield vs Time' dependencies. Although the results obtained for ball-milled scrap samples looked competitive with those for the composites with Wood's alloy, this alloying additive still has potential for application because of its low melting temperature. It was suspected that the alloy concentrations tested in this study may turn out to be excessive, leading to the opposite effects, and that the alloy portion can be substantially reduced. Therefore, further investigations on adjusting the content of Wood's alloy and ball-milling conditions should be performed in order to reveal the optimal composition and to elaborate usable material for hydrogen production. The tested temperature interval can be extended as well.

It should be noted, however, that Pb and Cd can produce some toxic influences, and this is the reason why Wood's alloy should be handled in a proper manner. Thus, the alloyed hydroreactive materials are not intended for implementation everywhere by everyone. After optimization of the alloy content, a toxicity characteristic leaching procedure (TCLP) should be performed for the reaction products in order to avoid contamination with heavy metals. Depending on the results, the product can be disposed of with common waste, undergo decontamination procedures, or be utilized in some other way. The produced hydrogen should be tested for corrosive or toxic properties as well to determine whether its treatment is required. However, Wood's alloy is a commercial product and, at aircraft manufacturing plants, Mg-Al scrap is available in abundance. Therefore, the said hydroreactive materials have potential for in situ production and conversion into hydrogen to provide energy for balance-of-plant needs.

4. Conclusions

In the present study, the hydrogen production performances of different powder composites elaborated by ball milling were compared. It was shown that the samples with a salt additive had the finest particle size and provided fast reactions proceeding in the beginning of the process. However, after a longer reaction time, their hydrogen yields turned out to be considerably lower than those for the samples without salt, although for some time after the beginning, they reacted quite slowly.

The observed effects resulted from the difference in the powder particle sizes (apparently affecting specific surface area), crystal lattice imperfections accumulated during ball milling, which enhance pitting corrosion in the presence of Cl^- ions, morphology of the solid reaction product particles covering the powders, and the contradicting effects from the potential formation of reaction-enhancing microgalvanic cells and reaction-hindering crystal grain screening compounds of the alloy and metal scrap components.

It was demonstrated that, in the reaction beginning, the sample with 20 wt.% KCl rapidly reacted with the aqueous solution because it was composed of the finest powder particles (and, subsequently, had large specific surface area). However, after nearly 20 min, the reaction was largely slowed down due to the formation of a 'dense' and 'compact' reaction product layer covering the particles. The investigated reaction product morphology for this sample suggested this version, while for other reaction product samples, the aggregations of large-sized 'flakes' covered either the entire particle surface or a large part.

The hydrogen yield for the sample with 20 wt.% Wood's alloy was one of the highest for 4 h of experiment. This additive was expected to create 'microgalvanic cells' of Mg-Al scrap and Wood's alloy heavy metal components. The said 'microgalvanic cells' were intended to induce anodic Mg dissolution in the conductive media, ensuring enhanced hydrogen production. However, the effect of Wood's alloy was negligible compared to the data obtained for the powder of the ball-milled scrap without additives. It is known that some compounds originating from alloying tend to locate along crystal grain boundaries. If their amounts are not exceeding a certain optimal value, they can promote corrosion of the base metal in the grains. However, extra amounts of some compounds or elements can lead to an opposite effect of screening grains from liquid. For the present study, it was assumed that the obtained results were associated with alloy contents largely exceeding the optimal value.

The sample composed of 10 wt.% KCl and 10 wt.% Wood's alloy was expected to demonstrate the best hydrogen production performance in terms of yield and reaction rate due to the synergetic effect of particle size reduction and the formation of 'galvanic couples'. However, the results for this composition fell between those obtained for the two samples discussed above.

For the ball-milled sample without additives, it was established that it combined a number of reaction-enhancing properties of other powder samples. In particular, it did not contain compounds potentially screening base metal grains, it presumably accumulated many structure defects from ball milling, and its reaction product formed structures containing many clusters of large-sized 'flakes'.

Author Contributions: O.A.B., M.S.V. and A.I.K. designed the study; A.O.D. and G.N.A. wrote the introduction section; G.E.V. and A.V.G. performed analyses of the samples; O.A.B., A.O.D. and G.N.A. carried out the experiments; O.A.B. analyzed the results and wrote the rest of the manuscript; M.S.V. and A.I.K. supervised the study and edited the manuscript. All authors have read and agreed to the published version of the manuscript.

Funding: This work was funded by the Ministry of Science and Higher Education of the Russian Federation (State Assignment No. 075-01056-22-00).

Acknowledgments: This work was supported by the Ministry of Science and Higher Education of the Russian Federation (State Assignment No. 075-01056-22-00). (Recipients Buryakovskaya O.A., Vlaskin M.S., Valyano G.E., Grigorenko A.V., Ambaryan G.N., Dudoladov A.O.). This paper was prepared with the support of the 'RUDN University Program [RUDN University Strategic Academic Leadership Program]' (Recipient Kurbatova A.I.).

Conflicts of Interest: The authors declare no conflict of interest.

References

1. Fedotkina, O.; Gorbashko, E.; Vatolkina, N. Circular Economy in Russia: Drivers and Barriers for Waste Management Development. *Sustainability* **2019**, *11*, 5837. [CrossRef]
2. Kurbatova, A.; Abu-Qdais, H.A. Using Multi-Criteria Decision Analysis to Select Waste to Energy Technology for a Mega City: The Case of Moscow. *Sustainability* **2020**, *12*, 9828. [CrossRef]
3. Kuusiola, T.; Wierink, M.; Heiskanen, K. Comparison of Collection Schemes of Municipal Solid Waste Metallic Fraction: The Impacts on Global Warming Potential for the Case of the Helsinki Metropolitan Area, Finland. *Sustainability* **2012**, *4*, 2586–2610. [CrossRef]
4. Leite, R.; Amorim, M.; Rodrigues, M.; Neto, G.O. Overcoming Barriers for Adopting Cleaner Production: A Case Study in Brazilian Small Metal-Mechanic Companies. *Sustainability* **2019**, *11*, 4808. [CrossRef]
5. Ouyang, L.; Jiang, J.; Chen, K.; Zhu, M.; Liu, Z. Hydrogen Production via Hydrolysis and Alcoholysis of Light Metal-Based Materials: A Review. *Nano-Micro Lett.* **2021**, *13*, 134. [CrossRef]
6. Kara, S.; Erdem, S.; Lezcano, R.A. MgO-Based Cementitious Composites for Sustainable and Energy Efficient Building Design. *Sustainability* **2021**, *13*, 9188. [CrossRef]
7. Grigorenko, A.; Vlaskin, M. Densification of Porous Aluminum Oxide Powder by Plasma Arc Treatment. *IOP Conf. Ser. Mater. Sci. Eng.* **2018**, *381*, 012050. [CrossRef]
8. Silveira, N.C.G.; Martins, M.L.F.; Bezerra, A.C.S.; Araújo, F.G.S. Red Mud from the Aluminium Industry: Production, Characteristics, and Alternative Applications in Construction Materials—A Review. *Sustainability* **2021**, *13*, 12741. [CrossRef]
9. Ercoli, R.; Orlando, A.; Borrini, D.; Tassi, F.; Bicocchi, G.; Renzulli, A. Hydrogen-Rich Gas Produced by the Chemical Neutralization of Reactive By-Products from the Screening Processes of the Secondary Aluminum Industry. *Sustainability* **2021**, *13*, 12261. [CrossRef]
10. Yam, B.J.Y.; Le, D.K.; Do, N.H.; Nguyen, P.T.T.; Thai, Q.B.; Phan-Thien, N.; Duong, H.M. Recycling of magnesium waste into magnesium hydroxide aerogels. *J. Environ. Chem. Eng.* **2020**, *8*, 104101. [CrossRef]
11. Oslanec, P.; Iždinský, K.; Simančík, F. Possibilities of magnesium recycling. *Mater. Sci. Technol.* **2008**, *4*, 83–88.
12. Viswanadhapalli, B.; Bupesh Raja, V.K. Application of Magnesium Alloys in Automotive Industry—A Review. In *International Conference on Emerging Current Trends in Computing and Expert Technology*; Hemanth, D.J., Kumar, V.D.A., Malathi, S., Castillo, O., Patrut, B., Eds.; Springer International Publishing: Cham, Switzerland, 2020; pp. 519–531.
13. Li, Z.; Wang, Q.; Luo, A.A.; Peng, L.; Zhang, P. Fatigue behavior and life prediction of cast magnesium alloys. *Mater. Sci. Eng. A* **2015**, *647*, 113–126. [CrossRef]
14. Mukhina, I.Y.; Trofimov, N.V.; Leonov, A.A.; Rostovtseva, A.S. Development of Resource-Saving Technological Processes in the Metallurgy of Magnesium. *Russ. Metall.* **2021**, *2021*, 1394–1401. [CrossRef]
15. Uan, J.-Y.; Cho, C.-Y.; Liu, K.-T. Generation of hydrogen from magnesium alloy scraps catalyzed by platinum-coated titanium net in NaCl aqueous solution. *Int. J. Hydrogen Energy* **2007**, *32*, 2337–2343. [CrossRef]
16. Uan, J.-Y.; Yu, S.-H.; Lin, M.-C.; Chen, L.-F.; Lin, H.-I. Evolution of hydrogen from magnesium alloy scraps in citric acid-added seawater without catalyst. *Int. J. Hydrogen Energy* **2009**, *34*, 6137–6142. [CrossRef]
17. Uan, J.-Y.; Lin, M.-C.; Cho, C.-Y.; Liu, K.-T.; Lin, H.-I. Producing hydrogen in an aqueous NaCl solution by the hydrolysis of metallic couples of low-grade magnesium scrap and noble metal net. *Int. J. Hydrogen Energy* **2009**, *34*, 1677–1687. [CrossRef]
18. Yu, S.-H.; Uan, J.-Y.; Hsu, T.-L. Effects of concentrations of NaCl and organic acid on generation of hydrogen from magnesium metal scrap. *Int. J. Hydrogen Energy* **2012**, *37*, 3033–3040. [CrossRef]
19. Setiani, P.; Watanabe, N.; Sondari, R.R.; Tsuchiya, N. Mechanisms and kinetic model of hydrogen production in the hydrothermal treatment of waste aluminum. *Mater. Renew. Sustain. Energy* **2018**, *7*, 10. [CrossRef]
20. Buryakovskaya, O.A.; Meshkov, E.A.; Vlaskin, M.S.; Shkolnokov, E.I.; Zhuk, A.Z. Utilization of Aluminum Waste with Hydrogen and Heat Generation. *IOP Conf. Ser. Mater. Sci. Eng.* **2017**, *250*, 012007. [CrossRef]
21. Ambaryan, G.N.; Vlaskin, M.S.; Dudoladov, A.O.; Meshkov, E.A.; Zhuk, A.Z.; Shkolnikov, E.I. Hydrogen generation by oxidation of coarse aluminum in low content alkali aqueous solution under intensive mixing. *Int. J. Hydrogen Energy* **2016**, *41*, 17216–17224. [CrossRef]
22. Ambaryan, G.N.; Vlaskin, M.S.; Zhuk, A.Z.; Shkol'nikov, E.I. Preparation of High-Purity Aluminum Oxide via Mechanochemical Oxidation of Aluminum in a 0.1 M KOH Solution, Followed by Chemical and Heat Treatments. *Inorg. Mater.* **2019**, *55*, 244–255. [CrossRef]

23. Hiraki, T.; Yamauchi, S.; Iida, M.; Uesugi, H.; Akiyama, T. Process for Recycling Waste Aluminum with Generation of High-Pressure Hydrogen. *Environ. Sci. Technol.* **2007**, *41*, 4454–4457. [CrossRef] [PubMed]
24. Xiao, F.; Yang, R.; Liu, Z. Active aluminum composites and their hydrogen generation via hydrolysis reaction: A review. *Int. J. Hydrogen Energy* **2021**, *47*, 365–386. [CrossRef]
25. Grosjean, M.H.; Zidoune, M.; Roué, L. Hydrogen production from highly corroding Mg-based materials elaborated by ball milling. *J. Alloys Compd.* **2005**, *404–406*, 712–715. [CrossRef]
26. Grosjean, M.H.; Zidoune, M.; Roué, L.; Huot, J.Y. Hydrogen production via hydrolysis reaction from ball-milled Mg-based materials. *Int. J. Hydrogen Energy* **2006**, *31*, 109–119. [CrossRef]
27. Zou, M.-S.; Yang, R.-J.; Guo, X.-Y.; Huang, H.-T.; He, J.-Y.; Zhang, P. The preparation of Mg-based hydro-reactive materials and their reactive properties in seawater. *Int. J. Hydrogen Energy* **2011**, *36*, 6478–6483. [CrossRef]
28. Wang, C.; Yang, T.; Liu, Y.; Ruan, J.; Yang, S.; Liu, X. Hydrogen generation by the hydrolysis of magnesium–aluminum–iron material in aqueous solutions. *Int. J. Hydrogen Energy* **2014**, *39*, 10843–10852. [CrossRef]
29. Zou, M.-S.; Guo, X.-Y.; Huang, H.-T.; Yang, R.-J.; Zhang, P. Preparation and characterization of hydro-reactive Mg–Al mechanical alloy materials for hydrogen production in seawater. *J. Power Sources* **2012**, *219*, 60–64. [CrossRef]
30. Sevastyanova, L.G.; Genchel, V.K.; Klyamkin, S.N.; Larionova, P.A.; Bulychev, B.M. Hydrogen generation by oxidation of "mechanical alloys" of magnesium with iron and copper in aqueous salt solutions. *Int. J. Hydrogen Energy* **2017**, *42*, 16961–16967. [CrossRef]
31. Kravchenko, O.V.; Sevastyanova, L.G.; Urvanov, S.A.; Bulychev, B.M. Formation of hydrogen from oxidation of Mg, Mg alloys and mixture with Ni, Co, Cu and Fe in aqueous salt solutions. *Int. J. Hydrogen Energy* **2014**, *39*, 5522–5527. [CrossRef]
32. Li, S.-L.; Lin, H.-M.; Uan, J.-Y. Production of an Mg/Mg$_2$Ni lamellar composite for generating H2 and the recycling of the post-H2 generation residue to nickel powder. *Int. J. Hydrogen Energy* **2013**, *38*, 13520–13528. [CrossRef]
33. Ha, H.-Y.; Kang, J.-Y.; Yang, J.; Yim, C.D.; You, B.S. Role of Sn in corrosion and passive behavior of extruded Mg-5 wt%Sn alloy. *Corros. Sci.* **2016**, *102*, 355–362. [CrossRef]
34. Li, S.-L.; Song, J.-M.; Uan, J.-Y. Mg-Mg$_{2X}$ (X = Cu, Sn) eutectic alloy for the Mg$_{2X}$ nano-lamellar compounds to catalyze hydrolysis reaction for H$_2$ generation and the recycling of pure X metals from the reaction wastes. *J. Alloys Compd.* **2019**, *772*, 489–498. [CrossRef]
35. Buryakovskaya, O.A.; Vlaskin, M.S.; Ryzhkova, S.S. Hydrogen production properties of magnesium and magnesium-based materials at low temperatures in reaction with aqueous solutions. *J. Alloys Compd.* **2019**, *785*, 136–145. [CrossRef]
36. Tan, W.; Yang, Y.; Fang, Y. Isothermal hydrogen production behavior and kinetics of bulk eutectic Mg–Ni-based alloys in NaCl solution. *J. Alloys Compd.* **2020**, *826*, 152363. [CrossRef]
37. Xiao, F.; Guo, Y.; Yang, R.; Li, J. Hydrogen generation from hydrolysis of activated magnesium/low-melting-point metals alloys. *Int. J. Hydrogen Energy* **2019**, *44*, 1366–1373. [CrossRef]
38. Oh, S.; Kim, M.; Eom, K.; Kyung, J.; Kim, D.; Cho, E.; Kwon, H. Design of Mg–Ni alloys for fast hydrogen generation from seawater and their application in polymer electrolyte membrane fuel cells. *Int. J. Hydrogen Energy* **2016**, *41*, 5296–5303. [CrossRef]
39. Hou, X.; Wang, Y.; Yang, Y.; Hu, R.; Yang, G.; Feng, L.; Suo, G.; Ye, X.; Zhang, L.; Shi, H.; et al. Enhanced hydrogen generation behaviors and hydrolysis thermodynamics of as-cast Mg–Ni–Ce magnesium-rich alloys in simulate seawater. *Int. J. Hydrogen Energy* **2019**, *44*, 24086–24097. [CrossRef]
40. Hou, X.; Wang, Y.; Yang, Y.; Hu, R.; Yang, G.; Feng, L.; Suo, G. Microstructure evolution and controlled hydrolytic hydrogen generation strategy of Mg-rich Mg-Ni-La ternary alloys. *Energy* **2019**, *188*, 116081. [CrossRef]
41. Kantürk Figen, A.; Coşkuner Filiz, B. Hydrogen production by the hydrolysis of milled waste magnesium scraps in nickel chloride solutions and nickel chloride added in Marmara Sea and Aegean Sea Water. *Int. J. Hydrogen Energy* **2015**, *40*, 16169–16177. [CrossRef]
42. Kantürk Figen, A.; Coşkuner, B.; Pişkin, S. Hydrogen generation from waste Mg based material in various saline solutions (NiCl2, CoCl2, CuCl2, FeCl3, MnCl2). *Int. J. Hydrogen Energy* **2015**, *40*, 7483–7489. [CrossRef]
43. Shetty, T.; Szpunar, J.A.; Faye, O.; Eduok, U. A comparative study of hydrogen generation by reaction of ball milled mixture of magnesium powder with two water-soluble salts (NaCl and KCl) in hot water. *Int. J. Hydrogen Energy* **2020**, *45*, 25890–25899. [CrossRef]
44. Liu, Y.; Wang, X.; Dong, Z.; Liu, H.; Li, S.; Ge, H.; Yan, M. Hydrogen generation from the hydrolysis of Mg powder ball-milled with AlCl3. *Energy* **2013**, *53*, 147–152. [CrossRef]
45. Razavi-Tousi, S.S.; Szpunar, J.A. Effect of addition of water-soluble salts on the hydrogen generation of aluminum in reaction with hot water. *J. Alloys Compd.* **2016**, *679*, 364–374. [CrossRef]
46. Razavi-Tousi, S.S.; Szpunar, J.A. Effect of structural evolution of aluminum powder during ball milling on hydrogen generation in aluminum–water reaction. *Int. J. Hydrogen Energy* **2013**, *38*, 795–806. [CrossRef]
47. Mahmoodi, K.; Alinejad, B. Enhancement of hydrogen generation rate in reaction of aluminum with water. *Int. J. Hydrogen Energy* **2010**, *35*, 5227–5232. [CrossRef]
48. Alinejad, B.; Mahmoodi, K. A novel method for generating hydrogen by hydrolysis of highly activated aluminum nanoparticles in pure water. *Int. J. Hydrogen Energy* **2009**, *34*, 7934–7938. [CrossRef]
49. Al Bacha, S.; Awad, A.S.; El Asmar, E.; Tayeh, T.; Bobet, J.L.; Nakhl, M.; Zakhour, M. Hydrogen generation via hydrolysis of ball milled WE43 magnesium waste. *Int. J. Hydrogen Energy* **2019**, *44*, 17515–17524. [CrossRef]

50. Al Bacha, S.; Pighin, S.A.; Urretavizcaya, G.; Zakhour, M.; Castro, F.J.; Nakhl, M.; Bobet, J.L. Hydrogen generation from ball milled Mg alloy waste by hydrolysis reaction. *J. Power Sources* **2020**, *479*, 228711. [CrossRef]
51. Al Bacha, S.; Pighin, S.A.; Urretavizcaya, G.; Zakhour, M.; Nakhl, M.; Castro, F.J.; Bobet, J.L. Effect of ball milling strategy (milling device for scaling-up) on the hydrolysis performance of Mg alloy waste. *Int. J. Hydrogen Energy* **2020**, *45*, 20883–20893. [CrossRef]
52. Galwey, A.K.; Brown, M.E. *Handbook of Thermal Analysis and Calorimetry*; Elsevier Science B.V.: Amsterdam, The Netherlands, 1998; Volume 1.
53. Zhu, T.; Chen, Z.W.; Gao, W. Effect of cooling conditions during casting on fraction of β-$Mg_{17}Al_{12}$ in Mg–9Al–1Zn cast alloy. *J. Alloys Compd.* **2010**, *501*, 291–296. [CrossRef]
54. Lashko, N.F.; Morozova, G.I. *Some Issues on the Alloying and Phase Composition of Magnesium Alloys*; All-Russia Institute of Aircraft Materials (VIAM): Moscow, Russia, 1974.
55. Rosenband, V.; Gany, A. Application of activated aluminum powder for generation of hydrogen from water. *Int. J. Hydrogen Energy* **2010**, *35*, 10898–10904. [CrossRef]
56. Yavor, Y.; Goroshin, S.; Bergthorson, J.M.; Frost, D.L.; Stowe, R.; Ringuette, S. Enhanced hydrogen generation from aluminum–water reactions. *Int. J. Hydrogen Energy* **2013**, *38*, 14992–15002. [CrossRef]
57. Buryakovskaya, O.A.; Vlaskin, M.S.; Grigorenko, A.V. Effect of Thermal Treatment of Aluminum Core-Shell Particles on Their Oxidation Kinetics in Water for Hydrogen Production. *Materials* **2021**, *14*, 6493. [CrossRef] [PubMed]
58. du Preez, S.P.; Bessarabov, D.G. Hydrogen generation of mechanochemically activated AlBiIn composites. *Int. J. Hydrogen Energy* **2017**, *42*, 16589–16602. [CrossRef]
59. Zhang, F.; Edalati, K.; Arita, M.; Horita, Z. Fast hydrolysis and hydrogen generation on Al-Bi alloys and Al-Bi-C composites synthesized by high-pressure torsion. *Int. J. Hydrogen Energy* **2017**, *42*, 29121–29130. [CrossRef]
60. Al Bacha, S.; Farias, E.D.; Garrigue, P.; Zakhour, M.; Nakhl, M.; Bobet, J.L.; Zigah, D. Local enhancement of hydrogen production by the hydrolysis of $Mg_{17}Al_{12}$ with Mg "model" material. *J. Alloys Compd.* **2022**, *895*, 162560. [CrossRef]
61. He, T.; Chen, W.; Wang, W.; Ren, F.; Stock, H.-R. Effect of different Cu contents on the microstructure and hydrogen production of Al–Cu-Ga-In-Sn alloys for dissolvable materials. *J. Alloys Compd.* **2020**, *821*, 153489. [CrossRef]
62. Mohedano, M.; Blawert, C.; Yasakau, K.A.; Arrabal, R.; Matykina, E.; Mingo, B.; Scharnagl, N.; Ferreira, M.G.S.; Zheludkevich, M.L. Characterization and corrosion behavior of binary Mg-Ga alloys. *Mater. Charact.* **2017**, *128*, 85–99. [CrossRef]
63. Wang, X.; Chen, Z.; Guo, E.; Liu, X.; Kang, H.; Wang, T. The role of Ga in the microstructure, corrosion behavior and mechanical properties of as-extruded Mg–5Sn–xGa alloys. *J. Alloys Compd.* **2021**, *863*, 158762. [CrossRef]
64. Oh, S.; Cho, T.; Kim, M.; Lim, J.; Eom, K.; Kim, D.; Cho, E.; Kwon, H. Fabrication of Mg–Ni–Sn alloys for fast hydrogen generation in seawater. *Int. J. Hydrogen Energy* **2017**, *42*, 7761–7769. [CrossRef]

Article

Evaluation of Physio-Chemical Characteristics of Bio Fertilizer Produced from Organic Solid Waste Using Composting Bins

Aseel Najeeb Ajaweed [1,2,*], Fikrat M. Hassan [2] and Nadhem H. Hyder [3]

1. Department of Biology, College of Science, University of Baghdad, Baghdad 10071, Iraq
2. Department of Biology, College of Science for Women, University of Baghdad, Baghdad 10071, Iraq; fikrat@csw.uobaghdad.edu.iq
3. Department of Biotechnology, College of Science, University of Baghdad, Baghdad 10071, Iraq; nadhemwandawy@yahoo.com
* Correspondence: ssami@uob.edu.iq

Abstract: Background: The possibility of converting the organic fraction of municipal solid waste to mature compost using the composting bin method was studied. Nine distinct treatments were created by combining municipal solid waste (MSW) with animal waste (3:1, 2:1), poultry manure (3:1, 2:1), mixed waste (2:1:1), agricultural waste (dry leaves), biocont (Trichoderm hazarium), and humic acid. Weekly monitoring of temperature, pH, EC, organic matter (OM percent), and the C/N ratio was performed, and macronutrients (N, P, K) were measured. Trace elements, including heavy metals (Cd and Pb), were tested in the first and final weeks of maturity. **Results**: Temperatures in the first days of composting reached the thermophilic phase in MSW compost with animal and poultry manure between 55–60 °C, pH and EC (mS/cm) increased during the composting period in most composting bin treatments. Overall, organic matter (OM percent) and the C/N ratio decreased (10.27 to 18.9) as result of microbial activity during composting. Organic matter loss percent was less in treatments containing additives (biocont l humic acid) as well agricultural waste treatment. Composting bin treatments with animals and poultry showed higher K and P at the mature stage with an increase in micronutrients. Finally heavy metals were (2.25–4.20) mg/kg and (139–202) mg/kg for Cd and Pb respectively at maturation stage. **Conclusion**: Therefore, the results suggested that MSW could be composted in the compost bin method with animal and poultry manure. The physio-chemical parameters pH, Ec and C/N were within the acceptable standards. Heavy metals and micronutrients were under the limits of the USA standards. The significance of this study is that the compost bin may be used as a quick check to guarantee that the outputs of long-term public projects fulfill general sustainability requirements, increase ecosystem services, and mitigate the effect of municipal waste disposal on climate change particularly the hot climate regions.

Keywords: MSW; compost bin; organic matter; C/N; heavy metals

1. Introduction

Globally, urbanization and the continual growth of the human population have resulted in the development of massive amounts of trash. These waste streams have created a slew of environmental, social, and economic concerns, particularly in poor countries [1]. Each day, a substantial volume of municipal solid waste (MSW) is created and disposed of in landfills at a rate of less than 6% recycling and composting [2]. MSW management is without a doubt one of the most serious environmental concerns facing the world, especially in metropolitan areas. If not correctly treated, it may have a negative impact on the environment by creating a disagreeable odor, causing leachate, and emitting greenhouse gases. Landfilling is the primary method of disposing of MSW in the vast majority of nations globally, regardless of per capita wealth [3]. However, this treatment method has been criticized because of its high environmental impacts and incompatibility with the

concept of a circular bioeconomy [4]. Composting, as a less expensive and sophisticated method, may be used in place of landfills for organic waste recycling. Due to their high organic matter and mineral contents, municipal solid wastes may be utilized to improve soil fertility. Composting is an effective technique for generating a stabilized material that may be utilized in fields as a source of nutrients and soil conditioners [5]. Composting is an aerobic and exothermic process that is used globally to treat biodegradable trash and convert it into a soil conditioner and fertilizer. Additionally, it is the most ecologically beneficial technique of garbage disposal when compared to other processes, such as incineration, landfilling, and anaerobic digestion [6]. Composting has been shown to be an effective method of increasing crop productivity and soil quality. Composts include plant nutrients, particularly nitrogen and phosphorus, as well as organic substrates. As a result, it has the potential to influence a soil's physical, chemical, and biological quality [7]. Natural composting is a lengthy process, but land scarcity and vast amounts of MSW demand that these wastes be handled more rapidly. The amount and composition of organic matter in compost are critical in determining the product's quality [8,9]. Chemical parameters of composts, such as electrical conductivity (EC), pH, nutrient levels, organic matter, C:N ratio, and heavy metal concentration, have been utilized to determine their maturity and stability [10,11]. Three stages of composting were observed: the mesophilic stage increased temperature from ambient to 40 °C; the thermophilic stage characterized temperatures between 50 °C and 60 °C due to microbial activity during composting with the aid of passive or active aeration (turning composting bin) and moisture content between 40% and 60%). When the temperature reaches 60 °C, the breakdown process becomes sluggish, and microbial activity is restricted to thermotolerant bacteria. After 2–3 weeks, the temperature of the compost began to decrease and did not return to normal after turning. This is the cooling stage, during which the compost reaches stability after a period of time when it is resistant to decomposition. The factory for organic solid waste recycling in Al-Youssoufia City collects MSW collected from neighboring communities in southern Baghdad. The MSW comprised reusable organic materials as well as plastics and glass. The majority of this trash is disposed of in landfills. The present research aimed to convert the organic portion of MSW to biofertilizer using aerobic bin composting and to characterize the physicochemical properties of the compost. Urbanization and industrialization have created a slew of issues for garbage management. Pollution of the environment and garbage creation have a direct effect on biological diversity. Without proper treatment, haphazard discharge of solid waste into the environment may have negative consequences, resulting in contamination of all resources. Enhancing the pace and degree of degradation may provide considerable benefits in terms of optimizing the composting duration, quality, and production of the compost. This is accomplished by the addition of different chemicals to the composting process.

The purpose of this research is to determine the physicochemical properties of municipal solid trash in the presence of different additives, most of which are industrial waste products. The following goals were established for the research Composting of Solid Waste in Baghdad City Using Various Additives:

1. To discover and choose additives for organic waste stabilization.
2. To determine the viability of combining organic wastes for stabilization
3. To determine the physio-chemical parameters necessary for organic waste stabilization.
4. Performance evaluation of the method when coupled with various types of stabilized organic waste.
5. To compare the organic matter lose percent during composting duration.
6. To ascertain the presence and characteristics of heavy metals in solid waste.

2. Materials and Method

2.1. Feedstock and Composting Materials Used in the Composting Bin Methods

From September to December 2020, the composting was performed in the garden of the College of Science for Women at the University of Baghdad using the aerobic bin composting technique at an ambient, and compost maturity was monitored until March 2021.

Composting materials included MSW (which was previously crushed and sieved in the factory), which was the primary component of the composting process. The MSW was transported from a waste treatment and recycling facility in the district of the al-Yusufiya factory, approximately 20 km south of Baghdad. The dried animal and poultry manure utilized in the various mixtures was supplied by the Abu-Ghraib district's College of Agriculture Science.

Quality control and assurance has been provided for technical guidance in addition to the quality control and assurance of geosynthetics used to contain waste, as well as manufacturing quality assurance and control of these materials. It covers all types of waste containment facilities, including hazardous waste landfills and impoundments, municipal solid waste according to local and international standardization, and Starting with elaboration of sampling plan in accordance with the relevant guidelines and standards.

2.2. Design and Composting Methods

To construct compost heaps, composting bin scale composting was used, mostly via the use of composting containers [12,13]. Composting bin technique scale tests were performed in plastic containers (0.4 m × 0.4 m × 0.90 m). For aerobic composting, holes with a diameter of 1.5 cm were drilled at equal intervals in each container, approximately six rows on each side, to ensure aerobic composting. Composting bin s were physically turned twice a week for the first two weeks and then weekly until week 14. During the composting process, the moisture content of the compost should remain between 40% and 60%. Following the mixing of compost mixes, a thin layer of old compost was applied to the heaps to maintain warmth and moisture throughout the first stages of the composting process. Temperatures were taken daily for the first four weeks and thereafter weekly. For physio-chemical analysis, samples were gathered weekly after rotating the composting bin and then every two weeks, as previously stated. Samples were obtained from the center and margins of the organic waste composting bin and then combined in plastic bags to guarantee homogeneity of the sample for analysis. Composting took 14 weeks, but maturation and curing extended the process to 26 weeks. Each compost container was sieved using five-millimeter sieves. Weighing and removing nonbiodegradable particles from each compost treatment.

2.3. Experimental Design

Compost materials included the organic fraction of MSW, animal, poultry (chicken) manure and agricultural waste. Composting research employed the organic part of municipal solid waste as a starting material and animal, poultry, and agricultural waste as a bulking agent. The study involved nine main treatments consisting of the organic fraction of municipal solid waste mixed with different percentages of animal, poultry manure and agriculture waste. Biocont (*Trichoderm hazarium* 19×10^7 spore in 1 g) and potassium humate (K_2O 12%) as a source of carbon and nutrients [13] were added as separate treatments. Table 1 presents some of the physiochemical characteristics of the organic solid fraction (OSF) of municipal waste and the mixture of OSF with different bulking agents. The following treatments were used in the current study:

Composting bin 1: 100% organic solid waste (control).

Composting bin 2: organic solid waste: animal manure 3:1 (75:25) (w/w) %.

Composting bin 3: organic solid waste: poultry manure (chicken manure) 3:1 (75:25) (w/w) %.

Composting bin 4: organic solid waste: animal manure (2:1) (67:33) (w/w) %.

Composting bin 5: organic solid waste: poultry manure (2:1) (67:33) (w/w) %.

Composting bin 6: organic solid waste: agriculture waste 9:1 (90:10) (w/w) %

Composting bin 7: organic solid waste: animal manure: poultry manure (2:1:1) (50:25:25) (w/w) %.

Composting bin 8: organic solid waste: Bicont (100:2) ($w/w/v$) %.
Composting bin 9: organic solid waste: humic acid (100:2) ($w/w/v$) %.

2.4. Analysis of Physiochemical Characteristics

Temperature was monitored daily by a thermometer using a metal probe to make deep holes to reach the thermometer to a depth of 25–30 cm. All sample analyses were performed in the Biology Department College of Science and Laboratory of Agricultural Science Collage/University of Baghdad. Subsamples (500 g) were collected from each composting bin after mixing the components of the composting bins to obtain a representative sample every two weeks, air dried in an open container, crushed to sieve 2 mm and kept in polyethylene bags until analysis in the laboratory. The samples were kept in a refrigerator at 4 °C for physio-chemical analysis. A range of physiochemical parameters were determined, including the following:

2.4.1 Moisture was measured after turning and watering the composting bin samples, and 10 g of wet composted samples was weighed and dried in an oven at temperature 105 °C for 24 h.

2.4.2 The pH and electrical conductivity (EC) of a water extract were evaluated by diluting one part of the compost by volume with 10 parts distilled water at a ratio of 1:10 (w/v). The samples were shaken and allowed to precipitate before being filtered using the procedure described by [14], where both EC and pH has been calibrated with standard solutions.

2.4.2 The organic matter (OM) % and C% were calculated as the weight loss of the samples. Ten grams of composted sample was weighed and dried in an oven at 105 °C for 24 h and then ignited at 550 °C for 4 h. The difference in weight referred to as volatile substances and OM% was calculated according to the equations described by [15] as follows:

$$\mathrm{OM\%} = \frac{w1 - w2}{w1} * 100 \qquad (1)$$

where w1 is the dry weight after 105 °C and w2 is the weight after ignition at 550 °C.

$$\text{Organic carbon (C\%)} = \frac{\mathrm{OM\%}}{1.8} \qquad (2)$$

$$\text{Organic matter loss (OM\%)} = 100 - 100 \frac{X1[(100 - X2)]}{X2[(100 - X1)]} \qquad (3)$$

where X1 is an intail organic mater, and X2 is the final value of organic mater.

2.4.3 Macro-Nutrients (N,P,K) total nitrogen was determined by the modified Kjeldahl's method. The composted samples were digested in concentrated H_2SO_4 [16].

Total potassium was measured by the flame method using a flame photometer, and total phosphate was measured by a spectrophotometer at 882 nm by Jackson, 1973; After Prepare a series of Standard Solutions from the Stock Solution.

2.4.4 Mineral elements such as Cu, Zn, Mn and Fe as well as heavy metal concentrations (Cd and Pb) were determined by digestion of compost with concentrated H_2SO_4 and concentrated $HClO_4$ and atomic absorption spectrophotometry by Lindsay, W.1978.

3. Results and Discussion

3.1. Chemical Composition of Organic Wastes in Composting Bins

Municipal organic waste is one of the possible nutrient-rich organic leftovers that, when recycled, yields a useful and nutrient-dense product called compost.

The organic part of the trash was discovered (pH 6.7) and included a rather small amount of total nitrogen (1 percent). The organic carbon (C%) was 29.2%, with a C:N ratio of 29.272. The chemical features of OFSW (Table 1) indicated that it should be treated with additional nutrients to increase its value as an organic fertilizer. This would increase the microbial activity, stability, and maturity of the waste by composting. Provided a supply of nitrogen to maintain an optimal C/N ratio, as well as organic waste rich in N, P, and K nutrient [16–18]. Animal manure, chicken dung, and agricultural waste are all high in nitrogen and have been used to replenish heaps to remedy nitrogen depletion, as well as animal manure and poultry as additives or inoculums for compost treatment, which is beneficial for producing high-quality fertilizer or substrate hater 2015).

Table 1. Physio-chemical characteristics of raw and mixtures of composted materials.

Raw Materials and Their Mixture Used in the Composting Bins	pH	EC (ms/cm^{-1})	Organic Matter (OM)%	C (%)	N (%)	C:N (%)
Unmixed						
Bin 1: OSFW * (100%)	6.70	3.8	52.69	29.272	0.99	29.5
Mixtures						
Bin 2: OFSW+ Animal manure (3:1) (75% + 25% by weight)	6.47	4.100	62.182	34.55	1.18	29.42
Bin 3: OFSW + Poultry manure (3:1) (75% + 25% by weight)	6.73	4.375	63.56	35.310	1.435	24.61
Bin 4: OFSW + Animal manure (2:1) (66% + 33% by weight)	6.49	4.63	66.53	36.96	1.15	32.248
Bin 5: OFSW + Poultry manure (2:1) (66% + 33% by weight)	6.53	4.66	64.21	35.67	1.31	27.238
Bin 6: OFSW + Agriculture waste (9:1) (90% + 10% by weight)	6.90	4.03	54.193	30.11	1.05	28.713
Bin 7: OFSW + animal + poultry manure (2:1:1) (50:25:25) (w/w) %	6.85	4.47	68.65	38.14	1.7	22.433
Bin 8: OFSW + Bicont (100: 2) (w/w/v) %	6.8	4.475	49.48	27.49	1.08	25.457
Bin 9: OFSW + Humic fulvic acid (100: 2) (w/w/v) %	6.81	3.67	52.36	29.09	1.4	20.786
Standard values suitable for composting Source: Standard [19–21]	5.5–8.0	-	>20	30–40	>0.6	25–50:1

* Organic fraction of solid waste.

The present study's findings indicated that proportioning mixed wastes in the mixes in Table 1 resulted in a good C/N ratio in heaps 1–9, which varied between 20.786 and 32.248, as many studies have largely agreed.

3.2. Physical and Chemical Characteristics

3.2.1. The Effect of Temperature

The temperature pattern indicates the presence of microbes and the beginning of the composting process. Daily temperature readings for each bin were taken, and the weekly average was calculated. Refer to Figure 1. Temperatures increased in the majority of compost heaps to values between 55–60 °C in combination wastes with varying percentages of animal and poultry manure during the early stages of the composting process. Gradual temperature increases to the thermophilic stage (40–60 °C) over the three-week composting process, indicating microbial breakdown of organic solid waste. The temperature decreases after the sixth week and then stabilizes indicating that the organic part of garbage has decomposed into compost [22,23]. Due to extended maturation period temperature increased at last weeks of maturation lined with increased ambient temperature (27–29 °C) in March. Throughout the composting process, the moisture level was maintained between 40% and 60%.

3.2.2. Effect of pH

At the start of the composting process, the pH of MSW and its combinations with varying degrees of acidification was between 6.5 and 6.8. Reduced pH at the start of composting indicates the development of organic acids, which subsequently became more basic in all compost bins and reached more than 7 in all waste after two weeks (Figure 2). Gradual increases in pH throughout the composting process restrict fungal development that thrives in an acidic environment, while aeration holes and regular composting bin turning minimize CO_2 trapped in the empty space between compost particles [24]. At the curing stage, the pH value of compost bins is approximately 8 as a consequence of decreased microbial activity and organic acid generation agreed with study [25–27].

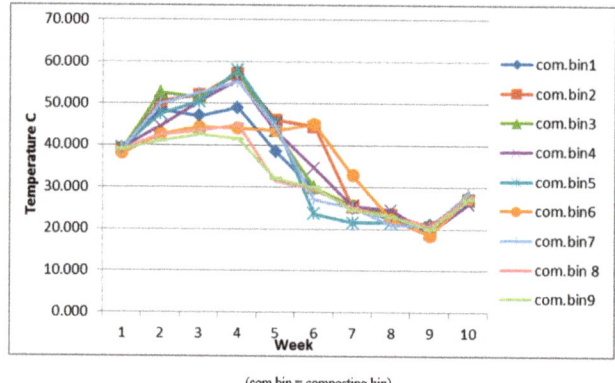

Figure 1. Temperature changes during the decomposition of organic waste in different composting bins.

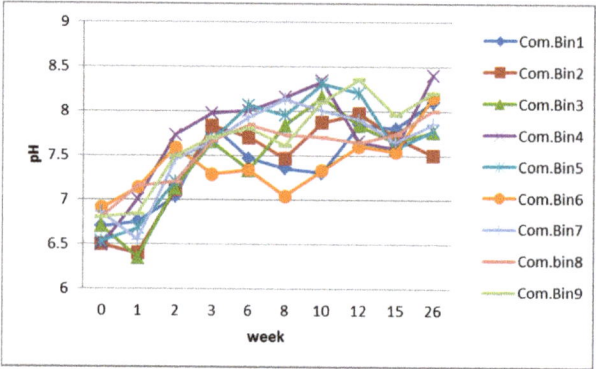

Figure 2. pH changes during the composting period in composting bins (1–9).

3.3. Electrical Conductivity (EC)

The EC value is critical to monitor during the composting process because it indicates the saltiness of the composted materials and their appropriateness for plant development. The present study discovered a considerable change in EC values throughout the composting process as a result of nutrient release. In comparison to the control and other compost combinations, the greatest EC values reported with compost mixtures of animal and poultry manure in the sixth week varied between 4.5–5.05 mS/cm (Figure 3). At the conclusion of the composting process (Figure 3), the EC value for MSW and combined organic wastes was substantially higher than the EC value for MSW alone.

Poultry chickens have a higher concentration of soluble salts because of the animal food they consume, but compost created from a combination of organic matter other than manure and plant material has a lower concentration of soluble salts [28]. EC values rise during organic matter degradation when inorganic compounds are formed, and the relative concentration of ions increases owing to waste mass loss [5]. The EC values for all compost combinations decreased as they matured as lined with study [29], while others showed an increase. There was a slight rise in EC values due to co-composting with organic additives [6,11,12,30]. As previously stated, the optimal EC value for developing plants is <4 dS m^{-1}.

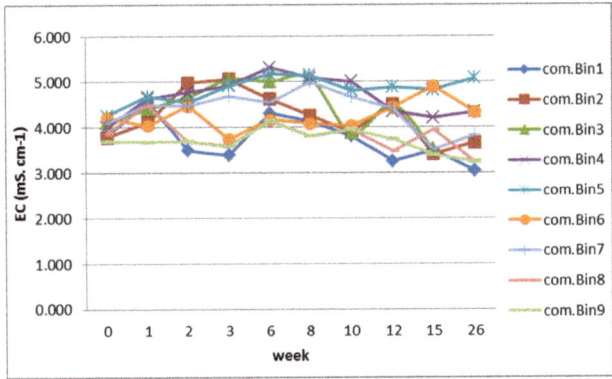

Figure 3. Effect of EC values on the decomposition of organic waste in different composting composting bin s.

3.4. Carbon:Nitrogen Ratio and Organic Matter

The C/N ratio is a critical component of the composting process. A high C/N ratio shows the existence of unutilized complex nitrogen, while a decreasing C:N ratio indicates the completion of the process (compost maturity). By and large, all waste mixtures exhibit a drop in organic carbon throughout the composting process as a result of organic matter breakdown. Figure 4. The C/N ratio for each composting bin ranged from (20.79 to 32.25) that percent which is closely to preferred C/N ratio stated in Doughtry 1998 the early stage of composting, decreased progressively as the substrate decomposed, and reached (10.27 to 18.9) percent during the mature stage, when it marginally decreased. Certain compost bins exhibit a slight rise in the C/N ratio, which might be due to ammonia volatilization under alkaline conditions with a low C/N ratio, as shown in compost bins 6 and 9 or bacteria extract nitrate or another nitrogenous molecule from compost, as well as compostingbin.2 resulted an increase in C /N in third week lined with [21] study when C/N increased during composting MSW alone at third week of composting process [30,31].

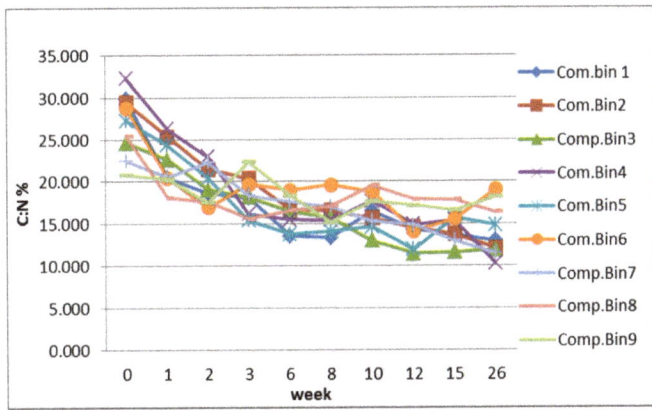

Figure 4. Changes in the C/N ratio during the decomposition of organic waste in different composting bins.

The drop in the C/N ratio might be the consequence of organic carbon being converted to carbon dioxide, followed by a decrease in the amount of organic acid. The reduction rate of C/N lowered between 6th–8th week agree with [32] study on food waste composting that TOC were more stable in 6th week. When the C/N ratio was less than 20, the compost was considered mature and may be utilized without limitation. Compost used in agriculture must have a C/N ratio between 12 and 18 [33,34].

The overall average organic matter (OM percent) content of the compost bin treatments varied between 41.93 and 53.3 during the composting period (Table 2). The organic matter loss was greater in compost bins 5 and 7 especially during thermophilic phase, where the average OM percent was considerably different from the MSW treatment. The addition of animal or poultry manure accelerated organic matter degradation in comparable rates between compost bin 2 and 3 but higher reduction when wastes mixed together in compostingbin 7 this due to higher organic matter content. The reduction rate was lower in wastes containing Compost bin 8 with Biocon. additives did not promote microbial activity in the same way that the Trichoderma hazarium spore suspension did in study [35]. The addition of potassium humate to MSW throughout the composting process had no effect on microbiological activity and increased degradation rate. Concerning compost quality, mature compost meets the standards for Eu guidelines (>15), and the CCQC recommends a minimum organic matter concentration of 25%.

Table 2. Organic matter (OM%) initial, mature, mean and organic matter losses in different treatments.

Composting Bin	OM% Initial	OM% Mature	OM% Mean	OM Loss
1	52.69	35.08	41.93	25.85
2	62.18	32.93	46.50	41.61
3	63.56	32.67	44.92	40.10
4	66.53	38.44	49.78	38.82
5	64.21	31.15	45.90	43.74
6	54.19	38.23	46.88	26.81
7	68.65	42.15	53.32	46.68
8	49.48	39.12	43.79	18.32
9	52.36	40.11	46.41	16.095
Week means	59.32	36.65	LSD composting-bin * week 2.352	

* Relation of avg. composting bin with weeks.

3.5. Effect of Nutrients (N, P and K) on the Composting Process

Generally, MSW mixed with animal manure and chicken poultry has an acceptable nitrogen percent for agricultural purposes. The variation between the initial and final values for compost due to microbial activity leads to an increase in the degradation of organic matter and loss of carbon in the form of CO_2 and the contribution of nitrogen-fixing bacteria, which are responsible for the increase in total nitrogen content at the mature stage [34]. After maturity, the present study found the highest nitrogen level in compost bins 4 and 7 (2.08% and 2.05%, respectively) and the lowest nitrogen content in composting bin 6 (1.120%), which is consistent with previous research on composted solid waste (Al-Turki, El-Hadidy, and Al-Romian 2013). Nitrogen loss was also detected at the mature stage as ammonia volatilization or immobilization processes, which were encouraged by high pH values (between 7.7 and 8.2), as evident in composting bins 5 and 9 [36,37].

Additionally, phosphorus and potassium are key minerals for plant growth. Generally, total phosphorous is represented as a percentage (%) of dry weight. The amount of P % in the initial stage of decomposition varied between 0.320 and 0.649% in all compostingbins (Table 3) and increased in the majority of composting bins during the final stage of the composting process as the compost cured. In composting bin 7 (organic solid waste:animal manure:poultry manure) (2:1:1), a slight increase in P percent was observed, ranging from 0.320 to 0.860%. Phosphorous loss (decrease) during composting may be attributed to phosphoric compound consumption in cell development or leachate. However, compost made from organic waste and poultry manure combined at a 2:1 ratio in compostingbin5 has a greater P percent content during the mature stage of decomposition. The current study's findings corroborated those of [38].

Table 3. Initial and final N, P and K contents in the composting process.

Composting Bins	Time (Week)	N%	P%	K%
1	0	0.990	0.410	0.565
	26	1.50	1.1	1.300
2	0	1.17	0.325	0.550
	26	1.520	0.695	1.510
3	0	1.435	0.320	0.565
	2	1.31	0.850	1.550
4	0	1.150	0.575	0.155
	26	2.08	1.390	1.370
5	0	1.310	0.649	0.610
	26	1.170	0.621	1.220
6	0	0.355	0.355	0.674
	26	1.120	1.200	0.775
7	0	1.700	0.642	1.190
	26	2.050	0.860	1.570
8	0	1.080	0.335	0.545
	26	1.330	0.520	0.705
9	0	1.400	0.624	0.800
	26	1.230	0.335	0.900
LSD composting bin		0.024	0.025	0.018

The total potassium concentration ranged between 0.155 and 1.19% in all composting bin s at the start of the composting process and concentrated toward the conclusion of the procedure, particularly for waste mixed with poultry manure. Additionally, the findings indicated that the compost generated from organic mixed materials in compostingbin7 (MSW:Animal: Poultry) (2:1:1) had a greater K level than the other treatments, owing to the waste including poultry manure, which is a source of K. In most composting composting bin s, the initial K percent was acceptable; this agrees with recent research that found animal manure had a lower K percent value than herbal plants and sugar cane [9,25].

3.6. The Effect of Heavy Metals on Composting Characteristics

The potential availability of Cd, Cu, Pb, and Zn increased with the age of compost due to the loss of organic matter from decomposition, and Fe was less toxic than other heavy metals because of the alkaline pH of compost [39]. The final values obtained for Cd, Pb, Cu, Zn, Mn and Fe were lower than the USA limits [40].

Based on the obtained results for the heavy metal content of composts (Figure 5), it can be concluded that the chemical analysis of heavy metals showed that Cd was higher in MSW and its mixtures in the beginning; after the composting process, the Cd decreased significantly, while it was lower in compost bins 1, 2 and 3 mixed with animal or poultry manure. However, the concentration of Cd largely decreased to lower levels after maturation, while the concentration of heavy metal pb was increased in all composting bins after maturation [41].

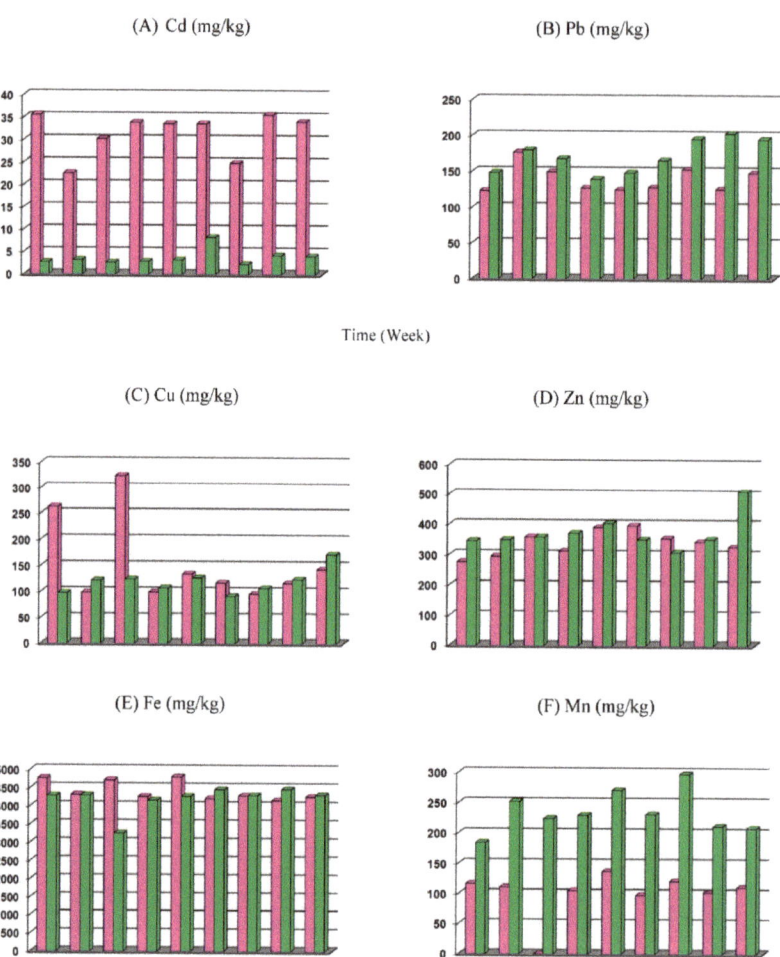

Figure 5. Heavy metal concentrations (mg/kg) at the beginning and end of maturation compost.

4. Conclusions

The current study was conducted to determine the effect of organic additives on the composting process of municipal solid waste collected from Baghdad's municipal solid waste processing unit. Numerous physicochemical elements influence the degradation process during composting, and the pH, EC, organic matter, total nitrogen, and C/N ratios of compost were all within the standards. Animal and poultry manure is an excellent source of bacteria and a bulking agent during the composting process. The macronutrient (N, P, and K) composition of manufactured compost is considered average when compared to compost made from animal and poultry manure and is not regarded as a cause of nutrient pollution in the environment. While the concentrations of Cu, Zn, Mn, and K increased during composting, the heavy metals and micronutrients remained below the permissible limits set by the USEPA for compost. Centralized waste management techniques become inadequate in an era of population boom. Demonstrating accessible alternatives and demonstrating the advantages of trash processing at the household level can motivate many individuals to take responsibility for their own rubbish.

Author Contributions: Conceptualization, A.N.A.; methodology, A.N.A.; software, A.N.A.; validation, A.N.A., F.M.H. and N.H.H.; formal analysis, A.N.A., F.M.H. and N.H.H.; investigation, A.N.A., F.M.H. and N.H.H.; resources, A.N.A.; writing—original draft preparation, A.N.A.; writing—review and editing, A.N.A., F.M.H. and N.H.H.; supervision, F.M.H. and N.H.H. All authors have read and agreed to the published version of the manuscript.

Funding: This research received no external funding.

Institutional Review Board Statement: Not applicable.

Informed Consent Statement: Not applicable.

Data Availability Statement: The data presented in this study are available on request from the authors.

Conflicts of Interest: The authors declare no conflict of interest.

References

1. Awasthi, M.K.; Pandey, A.K.; Khan, J.; Bundela, P.S.; Wong, J.W.; Selvam, A. Evaluation of thermophilic fungal consortium for organic municipal solid waste composting. *Bioresour. Technol.* **2014**, *168*, 214–221. [CrossRef] [PubMed]
2. Aghbashlo, M.; Tabatabaei, M.; Soltanian, S.; Ghanavati, H. Biopower and biofertilizer production from organic municipal solid waste: An exergoenvironmental analysis. *Renew. Energy* **2019**, *143*, 64–76. [CrossRef]
3. Salom, K.; Noor, T. Evaluation of waste sorting and recycling laboratory in Mahmudiya (case study). *J. Econ. Adm. Sci.* **2019**, *25*, 385–414.
4. Chew, K.W.; Chia, S.R.; Yen, H.-W.; Nomanbhay, S.; Ho, Y.-C.; Show, P.L. Transformation of Biomass Waste into Sustainable Organic Fertilizers. *Sustainability* **2019**, *11*, 2266. [CrossRef]
5. Zaman, A.U. A comprehensive study of the environmental and economic benefits of resource recovery from global waste management systems. *J. Clean. Prod.* **2016**, *124*, 41–50. [CrossRef]
6. Foereid, B. Nutrients Recovered from Organic Residues as Fertilizers: Challenges to Management and Research Methods. *World J. Agric. Soil Sci.* **2019**, *1*, 1–7. [CrossRef]
7. Bazrafshan, E.; Zarei, A.; Mostafapour, F.K.; Poormollae, N.; Mahmoodi, S.; Zazouli, M.A. Maturity and Stability Evaluation of Composted Municipal Solid Wastes. *Health Scope* **2016**, *5*, 1–9. [CrossRef]
8. Khan, M.; Sharif, M. Solubility Enhancement of Phosphorus from Rock Phosphate Through Composting with Poultry Litter. *Sarhad J. Agric.* **2012**, *28*, 415–420.
9. Zhao, S.; Liu, X.; Duo, L. Physical and Chemical Characterization of Municipal Solid Waste Compost in Different Particle Size Fractions. *Pol. J. Environ. Stud.* **2012**, *21*, 509–515.
10. Khater, E.G. Some Physical and Chemical Properties of Compost El-Sayed. *Int. J. Waste Resour.* **2015**, *5*, 172. [CrossRef]
11. Bernal, M.; Alburquerque, J.; Moral, R. Composting of animal manures and chemical criteria for compost maturity assessment. A review. *Bioresour. Technol.* **2009**, *100*, 5444–5453. [CrossRef] [PubMed]
12. Bustamante, M.; Paredes, C.; Marhuenda-Egea, F.C.; Espinosa, A.P.; Bernal, M.P.; Moral, R. Co-composting of distillery wastes with animal manures: Carbon and nitrogen transformations in the evaluation of compost stability. *Chemosphere* **2008**, *72*, 551–557. [CrossRef] [PubMed]
13. Storino, F.; Menéndez, S.; Muro, J.; Tejo, P.M.A.; Irigoyen, I. Effect of Feeding Regime on Composting in Bins. *Compost. Sci. Util.* **2017**, *25*, 71–81. [CrossRef]
14. Li, Y.; Fang, F.; Wei, J.; Wu, X.; Cui, R.; Li, G.; Zheng, F.; Tan, D. Humic acid fertilizer improved soil properties and soil microbial diversity of continuous cropping peanut: A three-year experiment. *Sci. Rep.* **2019**, *9*, 12014. [CrossRef] [PubMed]
15. Ujj, A.; Percsi, K.; Beres, A.; Aleksza, L.; Diaz, C.; Gyuricza, C.; Fogarassy, C. Analysis of quality of backyard compost and its potential utilization as a circular bio-waste source. *Appl. Sci.* **2021**, *11*, 4392. [CrossRef]
16. Sharma, D.; Varma, V.S.; Yadav, K.D.; Kalamdhad, A.S. Evolution of chemical and biological characterization during agitated composting bin composting of flower waste. *Int. J. Recycl. Org. Waste Agric.* **2017**, *6*, 89–98. [CrossRef]
17. Mamo, M.; Kassa, H.; Ingale, L.; Dondeyne, S. Evaluation of compost quality from municipal solid waste integrated with organic additive in Mizan–Aman town, Southwest Ethiopia. *BMC Chem.* **2021**, *15*, 43. [CrossRef]
18. Bremner, J.M. Determination of nitrogen in soil by the Kjeldahl method. *J. Agric. Sci.* **1960**, *55*, 11–33. [CrossRef]
19. Ahmad, R.; Jilani, G.; Arshad, M.; Zahir, Z.A.; Khalid, A. Bio-conversion of organic wastes for their recycling in agriculture: An overview of perspectives and prospects. *Ann. Microbiol.* **2007**, *57*, 471–479. [CrossRef]
20. Zuccconi, F.; de Bertoldi, M. Compost specification for the production and characterization of compost from municipal solid waste. In *Compost Production, Quality and the Use*; Applied Science Publishing: Basel, Switzerland, 1987.
21. Yousefi, J.; Younesi, H.; Ghasempoury, S.M. Co-composting of Municipal Solid Waste with Sawdust: Improving Compost Quality. *Clean-Soil Air Water* **2013**, *41*, 185–194. [CrossRef]
22. Woh, C.; van Fan, Y.; Suan, L.; Idayu, I. A Review on Application of Microorganisms for Organic Waste Management. *Chem. Eng. Trans.* **2018**, *63*, 85–90. [CrossRef]

23. Kosseva, M.R.; Kent, C.A. *Thermophilic Aerobic Bioprocessing Technologies for Food Industry Wastes and Wastewater*, 1st ed.; Elsevier Inc.: Amsterdam, The Netherlands, 2013. [CrossRef]
24. Liu, H.T.; Cai, L. Effect of Sewage Sludge Addition on the Completion of Aerobic Composting of Thermally Hydrolyzed Kitchen Biogas Residue. *BioResources* **2014**, *9*, 4862–4872. [CrossRef]
25. Hemidat, S.; Jaar, M.; Nassour, A.; Nelles, M. Monitoring of Composting Process Parameters: A Case Study in Jordan. *Waste Biomass Valorization* **2018**, *9*, 2257–2274. [CrossRef]
26. Olani, D.D.; Sulaiman, H.; Leta, S. Evaluation of Composting and the Quality of Compost from the Source Separated Municipal Solid Waste. *J. Appl. Sci. Environ. Manag.* **2012**, *16*, 5–10.
27. Jara-Samaniego, J.; Murcia, M.D.P.; Bustamante, M.A.; Paredes, C.; Espinosa, A.P.; Gavilanes, I.; López, M.; Marhuenda-Egea, F.C.; Brito, H.; Moral, R. Development of organic fertilizers from food market waste and urban gardening by composting in Ecuador. *PLoS ONE* **2017**, *12*, e0181621. [CrossRef]
28. Al-Turki, A.; El-Hadidy, Y.; Al-Romian, F. Assessment of chemical properties of locally composts produced in saudi arabia composts locally produced. *Int. J. Curr. Res.* **2013**, *5*, 3571–3578.
29. Tibu, C.; Annang, T.Y.; Solomon, N.; Yirenya-Tawiah, D. Effect of the composting process on physicochemical properties and concentration of heavy metals in market waste with additive materials in the Ga West Municipality, Ghana. *Int. J. Recycl. Org. Waste Agric.* **2019**, *8*, 393–403. [CrossRef]
30. Tognetti, C.; Mazzarino, M.J.; Laos, F. Improving the quality of municipal organic waste compost. *Bioresour. Technol.* **2007**, *98*, 1067–1076. [CrossRef]
31. Doughtry, M. *Composting for Municipalities: Planning and Design Considerations*; Natural Resource, Agriculture, and Engineering Service (NRAES): College Park, MD, USA, 1998.
32. Chaher, N.E.H.; Chakchouk, M.; Nassour, A.; Nelles, M.; Hamdi, M. Potential of windrow food and green waste composting in Tunisia. *Environ. Sci. Pollut. Res.* **2021**, *28*, 46540–46552. [CrossRef]
33. Acatrinei-Însurățelu, O.; Buftia, G.; Lazăr, I.M.; Rusu, L. Aerobic composting of mixing sewage sludge with green waste from lawn grass. *Environ. Eng. Manag. J.* **2019**, *18*, 1789–1798. [CrossRef]
34. Azim, K.; Soudi, B.; Boukhari, S.; Perissol, C.; Roussos, S.; Alami, I.T. Composting parameters and compost quality: A literature review. *Org. Agric.* **2017**, *8*, 141–158. [CrossRef]
35. Bari, M.A.; Begum, M.F.; Sarker, K.K.; Rahman, M.A.; Kabir, A.H.; Alam, M.F. Mode of action of *Trichoderma* spp. on organic solid waste for bioconversion. *Plant Environ. Dev.* **2007**, *1*, 61–66.
36. Altieri, R.; Esposito, A. Evaluation of the fertilizing effect of olive mill waste compost in short-term crops. *Int. Biodeterior. Biodegrad.* **2010**, *64*, 124–128. [CrossRef]
37. Gómez-Brandón, M.; Lazcano, C.; Domínguez, J. The evaluation of stability and maturity during the composting of cattle manure. *Chemosphere* **2008**, *70*, 436–444. [CrossRef] [PubMed]
38. Hussian, A.S.; Hyder, N.H.; Braesam, T.H.; Natheer, A.M. Recycling of Organic Solid Wastes of Cities to Biofertilizer Using Natural Raw Materials. *J. Al-Nahrain Univ. Sci.* **2016**, *19*, 140–155. [CrossRef]
39. Hooda, P.S. *Trace Elements in Soils*; Wiley Online Library: Hoboken, NJ, USA, 2010. [CrossRef]
40. Brinton, W.F. *Compost Quality Standards & Guidelines*; New York State Association of Recyclers: New York, NY, USA, 2000.
41. Al-Saedi, Z.Z.; Ibrahim, J.A.K. Evaluation of Heavy Metals Content in Simulated Solid Waste Food Compost. *J. Eng.* **2019**, *25*, 62–75. [CrossRef]

Article

Identification and Evaluation of Determining Factors and Actors in the Management and Use of Biosolids through Prospective Analysis (MicMac and Mactor) and Social Networks

Camilo Venegas [1,*], Andrea C. Sánchez-Alfonso [2], Fidson-Juarismy Vesga [1], Alison Martín [3], Crispín Celis-Zambrano [3,*] and Mauricio González Mendez [4]

[1] Department of Microbiology, Grupo de Biotecnología ambiental e industrial (GBAI), Laboratorio Calidad Microbiológica de Aguas y Lodos (CMAL), Pontificia Universidad Javeriana, Carrera 7 No. 43-82, Bogotá 110231, Colombia; vesga.f@javeriana.edu.co

[2] Corporación Autónoma Regional de Cundinamarca, Avenida Calle 24 (Esperanza) # 60-50, Centro Empresarial Gran Estación, Costado Esfera—Pisos 6-7, Bogotá 111321, Colombia; asanchez-a@javeriana.edu.co

[3] Department of Chemistry, Pontificia Universidad Javeriana, Carrera 7 No. 43-82, Bogotá 110231, Colombia; alison.martin@javeriana.edu.co

[4] School of Environmental and Rural Studies, Pontificia Universidad Javeriana, Transversal 4 No. 42-00, Piso 8°, Bogotá 110231, Colombia; gonzalez.alex@javeriana.edu.co

* Correspondence: c.venegas@javeriana.edu.co (C.V.); crispin.celis@javeriana.edu.co (C.C.-Z.)

Citation: Venegas, C.; Sánchez-Alfonso, A.C.; Vesga, F.-J.; Martín, A.; Celis-Zambrano, C.; González Mendez, M. Identification and Evaluation of Determining Factors and Actors in the Management and Use of Biosolids through Prospective Analysis (MicMac and Mactor) and Social Networks. *Sustainability* 2022, 14, 6840. https://doi.org/10.3390/su14116840

Academic Editors: Jose Navarro Pedreño, Caterina Picuno, Hani Abu-Qdais and Anna Kurbatova

Received: 29 April 2022
Accepted: 28 May 2022
Published: 3 June 2022

Publisher's Note: MDPI stays neutral with regard to jurisdictional claims in published maps and institutional affiliations.

Copyright: © 2022 by the authors. Licensee MDPI, Basel, Switzerland. This article is an open access article distributed under the terms and conditions of the Creative Commons Attribution (CC BY) license (https://creativecommons.org/licenses/by/4.0/).

Abstract: The reuse of biosolids in agriculture and its inclusion within the circular economy model requires evaluating and analyzing factors that intervene in its management. The objective of the study was to analyze those factors that influence the management and use of biosolids. Fifty-three actors were questioned, and their answers were analyzed using two prospective methods and Social Network Analysis (SNA) identifying between 14 and 19 variables. Six should be prioritized due to their criticality and potential in management and reuse scenarios. It was observed that the formulation of objectives, such as the improvement of infrastructure, creation of an institutional policy, and the establishment of definitions for the kinds of biosolids, are opposed by internal agents. Seven key actors and four to six determining agents were identified in the scenarios. The network of management and use of biosolids in agriculture presented low density (0.28) and the exclusive action of three key actors. Consequently, the participation of a greater number of better-connected actors is required to project networks with a higher density (between 0.49 and 0.57), facilitating the diffusion of information and the inclusion of new actors not previously contemplated. The application of prospective and SNA methodologies focused on biosolids allows the prioritization of determinants, the evaluation of the level of involvement and communication between actors, and other aspects that have not been considered previously in the management of WWTPs in Colombia.

Keywords: centrality measures; interest group; key variables; scenario-based planning; stakeholder analysis; waste management; wastewater treatment plants (WTTPs)

1. Introduction

The management of solid waste, as biosolids, is one of the main challenges for developing countries, due to the complexity and risk that this process represents in making decisions to develop an efficient, effective, and sustainable process [1,2]. Biosolids are a complex heterogeneous matrix and are the product of wastewater treatment. This product has received one or more stabilization treatments that would allow safe handling to be used [3–5]. Annually, around 2.5×10^7 to 6.0×10^7 tons of dry biosolids from Wastewater Treatment Plants (WWTP) are generated in the world [6]. In Colombia, lately, there has been an increase in the production of sludge and biosolids of domestic origin [7]. This is because of the improvement of the coverage, construction, expansion, and update rates of the infrastructure of wastewater treatment plants (WWTPs) in the country [8–10], although some impoverished areas still have low rates of coverage [11–14].

The use of biosolids in developed countries and some Latin American regions presents a greater preference for reuse in agricultural activities or direct application to the soil [4,15,16]. In Santiago de Chile, between 2009 and 2017 the use of biosolids in agriculture increased, taking advantage of close to 75% of the biosolids produced [17–19]. In the case of Brazil, in the decade from 2007 to 2017, in the state of Paraná, about 285,836 tons of biosolids were disposed of on agricultural land [20]. While in the case of Colombia the use of these residues in agriculture is almost null, despite having regulations that control this activity, such as Decree 1287 of 2014 [21] and the Colombian Technical Standard (NTC) 5167 (2011) [22].

Decree 1287 of 2017 establishes the determination and quantification of ten heavy metals, Arsenic (As), Cadmium (Cd), Copper (Cu), Chromium (Cr), Mercury (Hg), Molybdenum (Mo), Nickel (Ni), Lead (Pb), Selenium (Se) and Zinc (Zn), as well as thermotolerant coliforms, viable helminth eggs, *Salmonella* sp., enteric and somatic viruses, and phages as an alternative viral indicator. The permissible limits will vary depending on the quality of the biosolids reached (Class A or B), with class B being the one that will present greater restrictions of use compared to type A sludge [21]. The NTC 5167 standard determines the requirements and tests under which solid organic-mineral fertilizers or fertilizers, including biosolids, must be analyzed [22].

However, since 2002 research has been carried out in which this type of use has been evaluated [23]. The main activities of reuse currently carried out by the main WWTPs in Colombia are concentrated on restoring the soil in quarries, improving degraded soils, and preparing land for the entry of livestock, stabilizing slopes, seeding plants and shrubs [24–27]. Unfortunately, in the case of WWTPs in municipalities or other cities that have technical, operational, economic, and infrastructure limitations, it is difficult for them to obtain a Class B biosolid [21], which prevents its use.

In recent years, countries have focused on the acquisition of the circular economy model that includes the promotion of reuse, recycling, and recovery of waste. Thus, changing from the classic vision of the activities carried out by WWTPs towards an "ecologically sustainable system" can be influential and key within the process of wastewater treatment, by-products, use, and environmental sustainability [28–31]. Particularly in Colombia, through CONPES policy 4004 [32], it is projected that in the period from 2022 to 2025 biosolids will be included in the production cycle and their business opportunities will be defined. The use of biosolids in agriculture could become one of the most relevant options in the country, since it is one of the most sustainable and economical methods, especially for areas with technical and economic limitations [33–36]. In addition, taking this type of practice to the countryside, farmers, producers, and transformers would bring great benefits considering the vocation and agricultural capacity of the country [37].

To achieve this type of improvement, traditional management that has been implemented in Colombia must be put aside. This has been characterized by a focus on the evaluation of stabilization processes and the reduction in pathogens, the evaluation of alternatives for reuse or exploitation, and the updating of guidelines or regulations, among others [23]. For this reason, it is essential to identify, evaluate, and analyze other factors that may be affecting and limiting the management and use of biosolids. In research by Venegas et al. [23] through SWOT analysis, it was identified that the management of biosolids has been characterized by low cooperation, association, communication between actors and stakeholders, lack of updated and available information for decision-making, and a low level of use of this type of waste at a national level. Furthermore, the application of the SWOT methodology limited the consideration of other scenarios, because it was analyzed from a current vision or context and did not evaluate the focus and effectiveness of each of the proposed strategies, nor contrast the positions of favorability, disagreement, or neutrality of the different actors and institutions identified in this study [23].

According to the above, it is important to evaluate scenarios through methodologies that identify determinants and positions (neutrality/favorability/disagreement) of the different actors in the face of the proposed strategies, as well as the degrees of association of the networks of work and communication of one actor with another, regardless of power and influence, and their abilities to play the role of intermediaries, keys, and disseminators of information.

The application of different methodologies, such as social network analysis (SRA), and prospective analyses, including MicMac (Matrix of Cross Impacts Multiplication Applied to a Classification) which focuses on structural analysis and Mactor (Method, Actors, Objectives, Ratio on force) which evaluates stakeholders (actors) for the understanding of the management and use of biosolids in agriculture, would facilitate and strengthen the conversion from a linear model to a circular business model. The contribution and consideration of elements that have not been used or evaluated within the management and use of biosolids in Colombia and the possibility of being applied in sites with similar conditions and difficulties are factors that should be considered. This would open a framework of possibilities and considerations for the reuse of a material and its inclusion in an essential economic activity, such as the agricultural sector.

Prospective analysis is a method of reflection that allows the visualization of future scenarios about a specific subject. This encompasses making decisions that help to reduce possible risks and to pursue opportunities that others have not identified before. On one hand, the MicMac method aims to identify key, motor skills, and dependent variables that are typical of a system, distributed on a plane, and are classified according to their level of motor skills and dependence, based on the degree of influence perceived by the evaluators. On the other hand, from the key variables obtained by MicMac, the Mactor method allows the identification and evaluation of each of the actors along with their position (favorable, neutral, or disagreeing) in the face of a series of challenges or objectives proposed as transformative within the process of change or evolution of a system or scenario [38,39].

The analysis of social networks (SNA) is a tool that focuses on determining, identifying, and comparing the relationships within and between individuals, groups, and systems through the modeling or mapping of the different interactions that may be involved and related to "who knows who" and "who shares with whom" [40,41]. For the graphical representation of the analysis of the networks, nodes represent the actors or institutions, which are connected by edges, and these represent the relationships or flows existing between the nodes [42–44]. To identify which actors are essential in the network, the following metric evaluations are suggested [45,46]: degree of centrality, betweenness centrality, closeness of centrality, and density.

Consequently, the objectives of this study were: (I) to identify and analyze the factors that determine the management and use of biosolids from a WWTP in Colombia through prospective analysis (MicMac); (II) to evaluate stakeholders interested in the management and use of biosolids based on their influence/dependence on other actors and their positions of favorability, disagreement, or neutrality in the face of set objectives (Mactor); and (III) to map and analyze the type of social networks developed and projected by the different actors involved through the two systems.

2. Materials and Methods

2.1. Location and Characteristics of the WWTP

The evaluated WWTP is in the department of Boyacá, Colombia, which receives water collected by the sewerage network municipally from domestic, industrial, and rainwater wastewater. The treatment is an aerobic biological type; it is carried out using the activated sludge process in a Sequential Batch Reactor (SBR). The resulting sludge from the sedimentation process goes through a stabilization process and the biosolid is disposed of directly in the soil surrounding the WWTP infrastructure (Figure 1 and Table 1). In previous studies, it was determined that the quality of the biosolids produced by this WWTP makes their use difficult since the concentration of viable helminth eggs (VHE) [7] exceeds the limits established in Decree 1287 of 2014 [21] of the Ministry of the Environment of Colombia, without fulfilling the characteristics and parameters to classify the biosolids as class B.

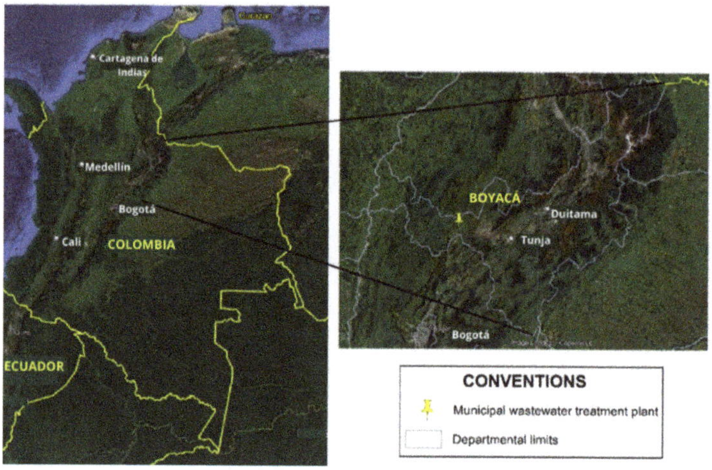

Figure 1. Location of the study area.

Table 1. Description of the WWTP treatment evaluated.

Treatment/Flow Treatment	Population Served	Water Line	Sludge Treatment	Type of Sludge Stabilization	Time of Treatment or Stabilization	Quantity of Treated Sludge Generated
SBR, AS/240 to 252 lps	~72.770 people	Pretreatment, primary, secondary, tertiary (UV light) treatment	Thickeners (polymers) and dewatering	Lime-treated	~1 month	~480 tons/year

~: Approximately, SBR: Sequencing Batch Reactor, AS: Activated sludge.

Average heavy metal concentrations in lime-treated biosolid for Cd, Cu, Cr, Mo, Ni, Pb, and Zn were 46.3, 61.4, 10.5, 4.4, 21.9, 17.2, and 1.1 mg/kg, respectively. Selenium was only detected in one sample, which presented a maximal concentration of 4.0 mg/kg. Furthermore, Ar (<4 mg/kg) and Hg (<0.5 mg/kg) were not detected. On the other hand, the levels of total, thermotolerant coliforms, and *E. coli* are 6.4, 5.9, and 5.4 Log10 CFU/g, respectively. For *Salmonella* spp., the values were 3.1 MPN/25 g, total helminth eggs 53.9 (HET)/4 g, and viable 19.9 (VHE)/4 g. In contrast, the concentration of viral indicators (Somatic Coliphages) corresponds to 5.6 Log10 PFU/g [7].

2.2. Identification of Actors in the Management and Use of Biosolids

The identification of actors was carried out at the national, departmental, and municipal levels, selecting those that were or could be related to the management practices and use of this type of waste. Subsequently, each of the interested parties was characterized according to their functions and roles [23] (Appendix A—Table A1). A survey designed through the Google forms© platform was sent to the different public and private entities that agreed to participate in the current study (Figure 2).

Fifty-three stakeholders participated, these belonging to the following different entities: WWTP of cities and municipalities, control or surveillance agencies (Corporación Autónoma Regional de Cundinamarca—CAR, Instituto Colombiano Agropecuario—ICA), farmers or associations, agroindustry, academia (Acad_), waste managers, national entities (Departamento Nacional de Planeación—DNP), Municipal and city WWTPs, and the

Economic and Agricultural Development area (Area de Desarrollo Económico y Agrícola—DEyA) of the municipality understudy (Appendix A—Table A1).

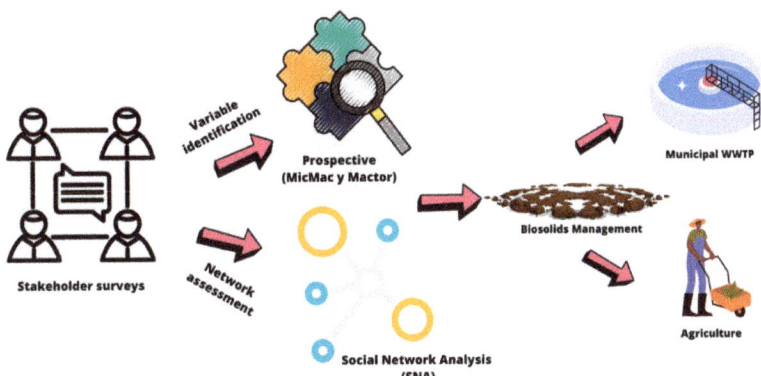

Figure 2. Methodological diagram.

2.3. Surveys and Private Interviews

The survey was conducted from January to July 2021. Its objective was to evaluate the incidence of a series of variables or determinants related to the management of sludges and biosolids in the WWTP and the use of these wastes in agriculture. Through the specific interviews, the objective was to determine and characterize the different actors regarding their influence/dependence and their position of agreement or disagreement in the face of a series of challenges posed. Finally, the type of network connections was determined based on the exchange of information and collaboration between actors (Figure 2).

2.4. Data Analyses

The different data collected from the surveys were consolidated in Microsoft Excel© (Version 2110) and the qualitative evaluations were categorized by establishing numerical scales for the MicMac software (Version 5.3.0) [47], and Mactor (Version 5.3.0) [47] analyses. Gephi (Version 0.9.2) software was used for SNA [48]. Acronyms were used, facilitating the visualization and understanding of the distribution of the variables (Table 2 and Figure 2).

2.4.1. MicMac Method: Variables Identification and Structural Analysis

The formulation and identification of the variables or factors of change were carried out through interviews, literature review, and the results of the SWOT analysis described in Venegas et al. [23], which focused on the evaluation of Colombian regulations and management with the same WWTP objective of this study. The following two questions were analyzed: (I) What are the factors that condition the management of biosolids? (II) What are the factors that determine its subsequent use in agriculture? A total of 14 and 19 variables were obtained, respectively, for each question, which was entered in the survey and interviews (Table 3), and evaluated using a numerical scale. This was followed by the consolidation of all the evaluations and the development of the MicMac methodology proposed by Godeth et al. [38,47].

Table 2. List of acronyms and definitions of the variables or stakeholders.

Short Form	Long Title	Short Form	Long Title	Short Form	Long Title	Short Form	Long Title
AA	Available area	ESPB	Departmental Public Utility Company	SSPD	Superintendency of Domiciliary Public Utilities	WWTP-OTD	Operational and Technical Division of the WWTP
Acad_	Academia	F/A	Farmes or Associations	Risk	Risk		
AI	Agro-industry	ICA	Colombian Agricultural Institute	SA	Stakeholder Articulation		
AQ	Affluent Quality	IEP	Institutional environmental policy	ST	Stabilization time		
AS	Agronomic Studies	Inf_	Information	TE	Treatment Efficiency		
BM	Business Model	Inst_	Institutionality	TK	Traditional Knowledge		
BT	Biosolid Type	IPU	Institutional Policy of Use	Tra_	Training		
CAR	Regional Autonomous Corporation of Cundinamarca	May_	Mayoralty	TSR	Treatment System Renewal		
Cert_	Certifications	MC	Municipal Council	WM_	Waste Managers		
Comm_	Community	MinA_	Department of Environment and Sustainable Development	WWTP(s)	Wastewater Treatment Plants		
Commun_	Communication	MinV_	Department of Housing, City and Territory	WWTP-EM	Environmental manager of the WWTP		
Cost	Cost	Ope_	Operability of the WWTP	WWTP-FA	Financial Area of the WWTP		
DEyA	Economic and Agricultural Development Area	Pers_	Persons	WWTP-Infr_	The infrastructure of the WWTP		
DNP	National Planning Department	PS	Product Satisfaction	WWTP-Manag	Manager of the WTTP		
EI	Environmental Incentives	Reg_	Regulations	WWTP-Op	Operators of the WTTP		

CAR: Corporación Autónoma Regional de Cundinamarca, DEyA: Área de Desarrollo Económico y Agropecuario, DNP: Departamento Nacional de Planeación, ESPB: Empresa Departamental de Servicios Públicos, MinA_: Ministerio de Ambiente y Desarrollo Sostenible, MinV: Ministerio de Vivienda, Ciudad y Territorio, SSPD: Superintendencia de Servicios Públicos Domiciliarios.

Furthermore, the double-entry matrix (structural analysis matrix) was used for the evaluation of the effect and interactions of the study variables. The qualification of each of the variables (Table 3) was carried out according to the following question: Is there a direct influence relationship between variable i and variable j? The rating was used according to the influence scale of 0 to 4 proposed by Godet et al. [38,47]. The assessment obtained was entered into the MicMac software [38,47]. Finally, the identification of subgroups of variables (environment variables, regulators, objective, keys, outcome, autonomous, determinants, and secondary levers) was inferred from their location and distribution in the influence and dependence plot.

Table 3. Variables or factors for the management of biosolids in the WWTP and their subsequent use in agriculture.

Short-Form	I	II	Definitions
AQ	X	X	Presence of organic and inorganic compounds in the water that enters the WWTP for treatment
AS		X	Studies to determine the type of soil and quantity of biosolids to be added according to the conditions and requirements of the soil and the type of crop
AA	X		Space available to the WWTP to carry out stabilization processes and temporary or final disposal of sludge and/or biosolids
BT		X	Class or type of biosolid generated by the WWTP for its use
BM		X	Definition and establishment of the form of distribution and destination of the profits of the stakeholders involved
Cert_		X	Obtaining the endorsement or authorization for the distribution of the stabilized product for use in the agricultural sector
Commun_		X	Disclosure of the management, results, and quality of the biosolids obtained
Comm_	X	X	Inclusion of the communities or group of people involved in the management of sludge and biosolids and in processes that lead to acceptance of reuse in agriculture.
Cost	X		Final treatment costs passed on to the aqueduct and sewer users
Cost		X	Costs of biosolid stabilization treatment to class A and B and compliance with decree 1287
EI	X		Incentives that the WWTP receive for the proper management and disposal of sludge and biosolids
Inf_	X	X	Availability of updated information on aspects related to the management of biosolids for both control entities and interested stakeholders
WWTP-Infr_	X		Current conditions of operation and proper functioning of the WWTP
IEP	X		Adoption and fulfillment of an institutional mission and vision focused on adequate management of biosolids
IPU		X	Adoption and fulfillment of a mission and vision by public service institutions focused on the proper use, control, and monitoring of biosolids in agriculture
Inst_	X	X	Presence and coordination of public and private entities to control compliance with current regulations related to sludge management and the use of biosolids.
Ope_	X	X	Correct operation of the WWTP complying with the times and other parameters established from the design, instructions given by the manufacturer, and operation manual
PS		X	Satisfaction of the people who use the product
Reg_	X	X	Compliance with the requirements or parameters indicated in the standards or decrees that regulate sludge management and the production, classification, and reuse of biosolids in agriculture
Risk		X	Detection and reporting of the presence of emerging persistent pollutants in biosolids
ST		X	Additional time is required for the product to exhibit Class A or B biosolids characteristics
SA		X	Linking public and private actors for the management, commercialization, and use of biosolids in agriculture
TK		X	The difference in perceptions between actors regarding the use of biosolids in agriculture
Tra_	X		Training of personnel for the treatment and control of the stabilization process of sludge and biosolids
TE	X	X	The efficiency of a team or a series of processes to obtain usable waste under Decree 1287 of 2014
TSR	X		Degree of updating or renovation of sewage and sludge treatment system

(I) factors that condition or affect the management of biosolids, (II) factors that condition or affect the use of biosolids in agriculture.

2.4.2. Mactor Method: Stakeholder Strategies Analysis

The different phases to carry out the identification and evaluation between actors and the analysis of the positions of disagreement/agreement with the challenges and objectives are described below. This is based on two types of rating scales that correspond to the influence between actors and the intensity of positioning against the proposed objectives.

Identification of Actors, Strategic Challenges, and Associated Objectives

The actors that could influence the 26 variables identified and listed in the MicMac methodology (Table 3). Based on the six key variables identified, six strategic challenges and 12 associated objectives were proposed to later expose them to each of the actors, to obtain and identify their position of neutrality, disagreement, or favorability (Table 4).

Table 4. Challenges and objectives associated with the key variables of the management and use of biosolids in the WWTP understudy.

	Challenges and Objectives of Biosolids Management in the WWTP Understudy	
Short-Form		Description of the Proposed Objectives
TE	•	Document and implement mechanisms for the control, monitoring, and verification of sludge production with its stabilization processes within the second half of 2021.
IEP	• • • •	Prepare and adopt an institutional policy for the company that operates the WWTP that allows the stabilization of sludge until achieving 10% class B biosolids. Define a policy for the use of biosolids within the company that operates the WWTP projected for the first half of 2022. Involve municipal institutions or entities in the study area in the institutional policy of use within the first semester of 2022. Initiate the dissemination and integration of the institutional policy built, with 30% of the officials of the PTAR operating company
WWTP-Infr_	• •	Implement the necessary adjustments for stabilization using drying beds together with the addition of lime in the next 12 months. Implement corrective maintenance for 50% of the equipment that has failures as of 2021 and that are currently operated manually.
	Challenges and objectives for the use of biosolids in agriculture	
BM	• •	By 2022, start conversations and activities with the different actors involved in the distribution and use of biosolids. Define the mechanism or form of distribution of biosolids to start with the use in agriculture in 2022.
Cert_	•	Request to the ICA the registration of producer and/or distributor of biosolids as agricultural input and obtain it before the end of 2022.
BT	• •	Define the type or class of biosolid that is chosen in agriculture for the year 2022. Establish monitoring mechanisms for 10% of the substances evaluated in other international regulations.

Assessment of Influences between Actors

After the evaluation of influence and dependence from zero to four, proposed by Godet et al. [38,47], the values obtained were consolidated according to the power relationships of one actor over another, directly or by a third party. Then the middle values were entered into the Mactor software [47] obtaining a representation of Actors × Actors according to their influence-dependence level, classifying the actors as determining, liaison, dominated, or autonomous.

Ranking of Each Actor According to Their Priorities by Objectives

The evaluation of the strength of positioning of each of the identified actors in the face of the previously proposed challenges was carried out through an assessment from zero to four together with the +/− signs that signify favorability and disagreement, allowing us to obtain a representation of Actors x Objectives [38].

2.4.3. Social Network Analysis (SNA)

Based on the evaluations obtained from the surveys carried out for the 53 stakeholders in which the level of work or communication with the different entities was determined, the average of each of the relationships was entered through adjacency matrices to the Gephi software [48] (Version 0.9.2). The three types of social networks obtained and analyzed corresponded to: (I) Level of communication and work of the WWTP, (II) Desired level of communication and work in the management of the WWTP, and (III) Desired level of communication and work for the use of biosolids in agriculture. The analysis metrics chosen were as follows:

Betweenness centrality: This value is used to determine the role of the actors that become bridges or links of interactions in the "middlemen" network. That is, it shows when a person is an intermediary between two other people in the same group who do not know each other [45,46,49]. Degree centrality: It is defined as the number of links that a node has. In a network, the actor with the highest degree of centrality is considered the key actor in the network. Closeness centrality: Represents the ability of a node to reach others through a reduced number of routes or paths and the possibility of efficiently disseminating information [45,46,49]. Density: degree of network connection, where a value of zero (0) represents a total disconnection, while a value of one (1) indicates all network actors are directly linked to each other information [45,46,49].

3. Results

3.1. Identification of the Factors That Determine the Management of Biosolids, MicMac

Fourteen variables were identified in the management of biosolids and sludges in the WWTP of the current study, which was plotted in the influence-dependence map (Figure 3A). As key variables, treatment efficiency (TE), infrastructure (WWTP-Infr_), and institu-tional environmental policy (IEP) were identified. For fulfilling the system improvement with the key variables, those should work together with the following objective variables: regulations (Reg_), institutional participation (Inst_), and design/use of environmental incentives (EI) (Figure 3A).

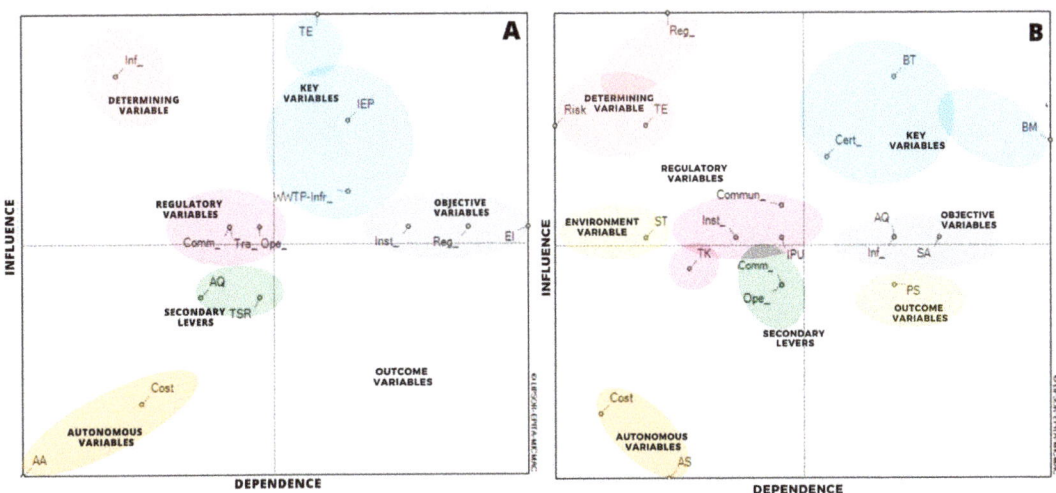

Figure 3. MicMac variables are distributed in the influence-dependence map. (**A**) Management of biosolids in the WWTP. (**B**) Agricultural biosolids utilization.

Other important variables which will impact the system performance and/or fulfillment of key variables are regulatory variables: communication (Commun_), operability of the WWTP (Ope_), and training (Tra_). However, treatment renewal system or biosolids stabilization variables (TSR) and affluent quality (AQ) could work together with the identified regulatory variables (Figure 3A). Information (Inf_) was identified among the determining variables as one that could act as a promoting or inhibitory factor depending on its evolution. Finally, cost (Cost) and available area (AA) were classified as autonomous variables, which are non-determining variables for the future of the system.

On the other hand, 19 variables that could influence the biosolids utilization were identified and distributed in the four quadrants of the influence-dependence map (Figure 3B). Key variables included: certification for distribution or commercialization (Cert_), biosolid type after stabilization (BT), and business model (BM), related to the product distribution and costs. Within the objective variables, information (Inf_), affluent quality (AQ), and stakeholders articulation (SA) were identified. The development of those variables could influence the positive evolution of the system (Figure 3B).

In addition, communication (Commun_), institutionality (Inst_), the institutional policy of use (IPU), and traditional knowledge (TK) were identified as regulatory variables. Community (Comm_) and operability of the WWTP (Ope_) were found as secondary lev-ers. Determining variables were risk (Risk), treatment efficiency (TE), and regulations (Reg_), the latter with a higher degree of influence (Figure 3B). Stabilization time (ST) was identified as an environmental variable, with few impacts on the improvement of the system (Figure 3B).

Cost (Cost) and agronomic studies (AS) were categorized as autonomous variables. Product satisfaction (PS) was found to be an outcome variable. Variables in this area give a descriptive indication of the system evolution, although it is not possible to approach them directly through those depending on the system (Figure 3B).

3.2. Evaluation of Stakeholders in the Management and Use of Biosolids, Mactor

Direct and indirect influences in the management of sludges and biosolids of the WWTP were determined for the 19 actors identified in the influence/dependence assessment (based on their interests in the development of the system) (Figure 4A). Determining, liaison, autonomous, and internal actors were identified.

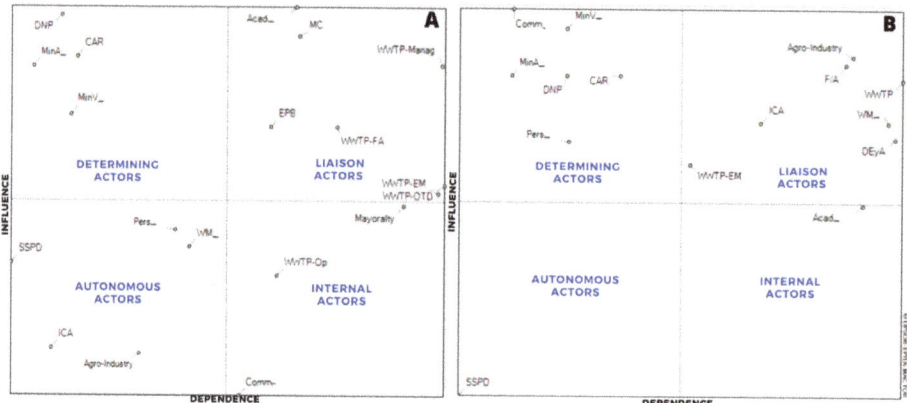

Figure 4. Results of the plane of influences and dependencies between actors of the management of sludge and biosolids in the studied WWTP and the use of biosolids in agriculture. Mactor influence/dependence map of the actors involved in the management of sludges and biosolids (**A**) and agricultural use of biosolids (**B**).

As shown in Figure 4A, determining actors included the National Planning Department (DNP), the Regional Autonomous Corporation of Cundinamarca (CAR), the Department of Housing, City and Territory (MinV_), and the Department of Environment and Sustainable Development (MinA_). The seven liaison actors were as follows: (I) Company in charge of WWTP management (WWTP-Mang), (II) Financial area of the WWTP (WWTP-FA), (III) Public Utilities Departamental Company—Boyacá (EPB), (IV) Academia (Acad_), (V) Municipal Council of the WWTP (MC), (VI) Environmental manager of the WWTP (WWTP-EM), and (VII) Technical and Operational Director (WWTP-OTD).

The actors with low influence and dependence, but that could act as secondary leverage, were also identified: Persons (Pers_), Waste Managers (WM), Superintendency of Domiciliary Public Utilities (SSPD), Colombian Agricultural Institute (ICA), and Agro-Industry (AI). Finally, Mayoralty (May_), WWTP operators (WWTP-Op), and Community (Comm_) were grouped as internal actors (Figure 4A).

Figure 4B shows the categorization and distribution of the 15 actors considered for the agricultural use of biosolids, distributed as follows: (I) Determining actors: community (Comm_), Department of Housing (MinV_), Department of Environment (MinA_),

DNP, CAR and Persons (Pers_); (II) Liaison actors: Agro-industry (AI), Farmers or Associations (F/A), Waste managers (WM_), ICA, Economic and Agricultural Development Area (DEyA), Environmental Manager of the WWTP (WWTP-EM) and WWTP; (III) Autonomous actor: Superintendency of Domiciliary Public Utilities (SSPD); (IV) Internal actor: Academia (Acad_).

Considering the objectives proposed in Table 4 and power relationships between different actors both in the management of biosolids at the WWTP and its use in agriculture, it is shown in Figure 5 that the strategic challenges present a favorable position. The values observed on the X-axis indicate the strength or level of commitment of the stakeholder towards the proposed objectives—the higher the level, the more affinity.

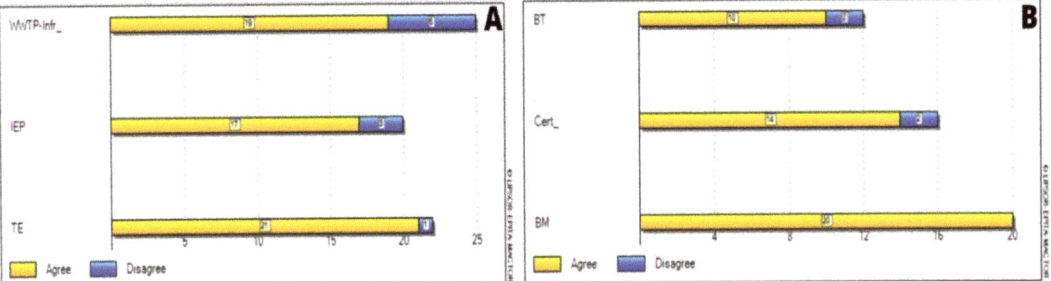

Figure 5. Histogram of stakeholder involvement on objectives. (**A**) WWTP management. (**B**) Agricultural use of biosolids.

The WWTP infrastructure (WWTP-Infr_) and the Institutional Environmental Policy (IEP), showed a lower level of commitment, and followed by Treatment Efficiency (TE) (Figure 5A). For the biosolids utilization scenario, the Business Model (BM) had a higher power relationship, while obtaining certifications (Cert_) and determining the Biosolids Type (BT) were not favorable for the study (Figure 5B).

3.3. Mapping of Social Networks Analysis (SNA) for the Management and Use of Biosolids

From the responses and ratings obtained from the stakeholders (actors) involved in both the management of the WWTP and the agricultural use of the biosolids, three network diagrams were obtained. Figure 6 represents the current communication and work state of the actors, the WWTP management, and biosolids use. Figures 7 and 8 show the desired level of communication and work for both internal WWTP management and agricultural use of biosolids, respectively.

According to Figure 6, the current management of biosolids carried out in the WWTPs of some cities and municipalities is focused on three key actors which belong to the nodes of WWTPs, waste managers (WM_) and CAR, in which the degrees of centrality were 20 and 17. On the other hand, the highest betweenness centrality level was obtained with the WWTPs (32.8), which makes them an intermediary actor in management. In contrast, the community (Comm_) and the waste managers (WM_) were identified as the closest actors to the others and acted as information disseminators, based on the closeness degree reached by each one (1.0 and 0.83).

Figure 7 showed that the central node belongs to WWTPs as well, although a higher number of key actors were involved in this network (WWTPs, CAR, Acad_, WWTP-EM, Comm_, WM_, and Pers_), indicating a stronger centrality degree (Degree: 19–14). In terms of closeness, the same key actors were identified, and the appearance of the Mayoralty is highlighted. However, a higher affinity is observed between CAR, Mayoralty, Environmental Manager of the WTTP (WTTP-EM), Academy, and the WWTPs, which could be connected to the other nodes without covering long distances within the same identified network. In contrast, it could be inferred that CAR and WWTPs could become liaison actors with the other stakeholders of the network, due to the values obtained for the betweenness degree (2.25 and 2.05).

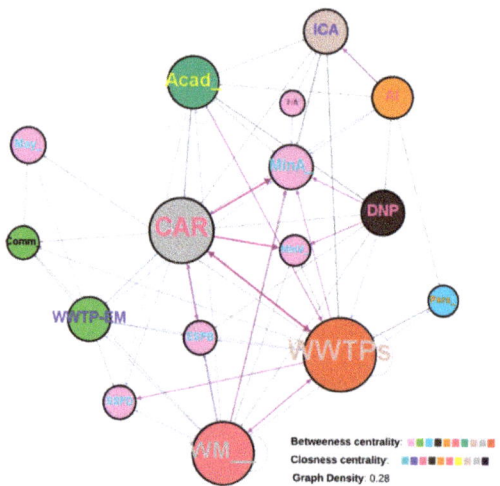

Figure 6. Communication and work level among different actors, management of the WWTP, and agricultural use of biosolids. The intensity of the colors of the edges represents stronger connections (weight). The larger the size of the nodes, the higher the degree level (degree). The colors of the node names represent the closeness centrality, with black being a higher closeness centrality value. Betweenness centrality is represented by the range of colors of the nodes from lilac to red, with the latter being the one with the highest valuation.

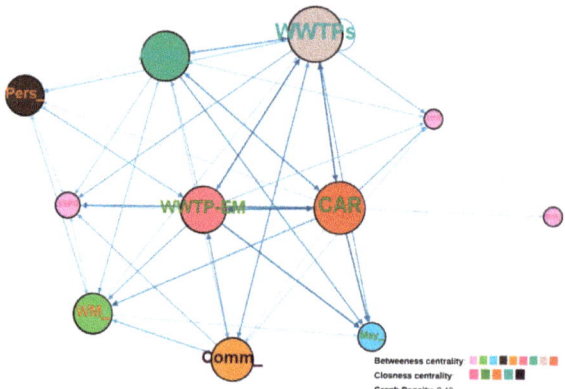

Figure 7. Desired levels of communication and work for the internal management of biosolids and sludges in the WWTP. The intensity of the colors of the edges represents stronger connections (weight). The larger the size of the nodes, the higher the degree level (degree). The colors of the node names represent the closeness centrality, with black being a higher closeness centrality value. Betweenness centrality is represented by the range of colors of the nodes from lilac to red, with the latter being the one with the highest valuation.

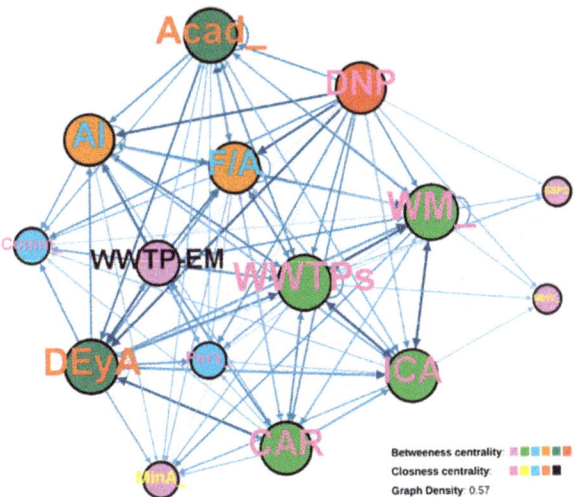

Figure 8. Desired level of communication and work between the different actors involved in the agricultural use of the biosolids. The intensity of the colors of the edges represents stronger connections (weight). The larger the size of the nodes, the higher the degree level (degree). The colors of the node names represent the closeness centrality, with black being a higher closeness centrality value. Betweenness centrality is represented by the range of colors of the nodes from lilac to red, with the latter being the one with the highest valuation.

Finally, Figure 8 indicates the desired level of engagement of the different actors for the utilization of the biosolids for agricultural purposes, which was focused on six key actors with a strength range between 25 and 27 (Degree) including: WWTPs, waste managers, academies, the Economic and Agricultural Development Area (DEyA), ICA, and CAR. The actors WWTPs, Waste Managers, ICA, and CAR, with a betweenness degree of 2 were also notable.

As for the density, scores close to 0 represent weak connections while scores close to 1 represent strong connections. In the case of the current management carried out in the WWTPs, a density score of 0.28 (Figure 6) was obtained, indicating a low and weak degree of integration between the existing actors in the management network. In contrast, desired networks of work and communication for both the management and utilization of biosolids revealed density scores of 0.49 and 0.57 in Figures 7 and 8, respectively, which were higher than the current density score of the WWTP management network. These results show opportunities for improvement in WWTP management and biosolids utilization supported by a better connection between actors in the proposed/desired networks, which could be enhanced with the participation of other stakeholders.

4. Discussion

4.1. Analysis of the Identification of the Factors That Determine the Management of Biosolids, MicMac

4.1.1. Management of the WWTP

According to the results obtained in the current study (Figure 3A), the high level of influence and dependence of the key variables (TE, EIP, and WWTP_Infr) could be explained due to the current conditions of operation and management of the WWTP. The lack of an organizational structure (policy, mission, vision, and objectives) focused on biosolids and sludges as well as the deficiency in the stabilization of sludges and conditions of the machinery of the WWTP could lead to the low quality of the obtained sludges interfering with further applications [50]. Thus, these key variables are considered by the different stakeholders as a priority and a key opportunity for the improvement of the WWTP.

On the other hand, objective (Inst_, EI, and Reg) and regulatory (Comm_, Ope, and TRA) variables were identified as essential tools that could act as enhancers to improve the management and operation of the WWTP. From those, it is important to highlight the impact of periodic training on all the employees and the community involved with the WWTP, as it leads to a better appropriation and knowledge of the system, resulting in a more proper operation of the processes carried out in the WWTP, with the final endpoint being an improvement in the decision-making of the WWTP [51,52]. Likewise, the environmental incentives (EI) for the management of sludges and biosolids became relevant in Colombia as governmental parties have encouraged companies to implement national policies which lead to better conversion and good use of biosolids and sludges [53–55].

In addition, secondary levers variables (TSR and AQ) could act as a complement for the system as these could impact both key and regulatory variables. Treatment system renewal (TSR) could favor the stabilization processes, nutrients' recovery, and enhancement of derived products, such as fertilizers obtained from the sludges [56]. Furthermore, affluent quality (AQ) has a direct impact on the biosolid types obtained after processing, which in some cases could increase the costs of the overall process depending on its quality, and the amount of trace elements, such as heavy metals and pesticides, among others [57–59], affecting about the 20–60% of the total budget of the WWTP [60,61]. Thus, AQ should be considered an essential variable that should be monitored in terms of Emerging Organic Compounds (EOC) and other residues present in the biosolids and sludges as a result of nearby industrial activities or other external contamination sources, even if it is not included in the regulations of most of the Latin American countries [21,62–64] except for Brazil [65–67].

Finally, the identification of "information" (Inf_) as a determining variable is crucial as the accessibility and availability of updated information could impact (positive or negative) the decision-making parties and planners for the management of the WWTP. Similarly, information should be available for the community, encouraging their participation and ensuring that the decision-making process is clear and sustained, increasing the effectiveness and efficacy of public policies and regulations [68,69].

4.1.2. Agricultural Use of Biosolids

In terms of the use of the biosolids (Figure 3B), biosolid type, business model, and certifications (BT, BM, and Cert_) were identified as key variables, which means that if those gain intervention, it will affect the overall performance of the system, likely resulting in an improvement in the use of biosolids (Figure 3B). Defining a business model (BM) and the articulation of public and private entities under the companionship of local governments favors access to the biological fertilizers market and direct application in the soil for agricultural practices [70,71]. Strategic alliances between different organizations could benefit the management and use of biosolids [72]. Obtaining certifications for the final products provide an advantage for the commercialization of fertilizers and direct use in the soil, as it represents an aggregated value and differential factor for the final product [73].

Actions between objective variables (AQ and Inf_) in this case represent a higher level of influence and dependence compared with the impact in the management of the WWTP in which those were identified as secondary and determining levers (Figure 3A). Interestingly, here stakeholders' articulation (SA) was identified as another objective variable in contrast with the WWTP management, as several direct and indirect actors are interested in biosolids utilization [36]. The articulation and integration of the different stakeholders in which responsibilities and functions are established will pave the way for the achievement of the objectives and challenges proposed for wastes management with an effective and sustainable process [74]. As an example of third parties' participation in Colombia, certain public and private companies are willing to provide consulting services for biosolids management [23], as well as current national policies willing to include WWTP by-products in the circular economy model [32].

On the other hand, secondary (Comm_ and Ope_) and regulatory (Commun_, Inst_, TK, and IPU) variables would help key and objective variables to work properly, as a clear institutional policy in conjunction with assertive communication will articulate individual efforts towards the same objective. In terms of the traditional knowledge (TK) of the community, and agreement with a previous study [23], it is a favorable position due to the interest in further applications of biosolids in agriculture for both the community and stakeholders.

The determining variables (Reg_, Risk, TE) become crucial factors for the system and could lead to opposite positions in the stakeholders and management of the biosolids [36,75–78]. One of the most important is risk perception/assessment, as the incorporation of biosolids in agriculture and soil could represent a hazard for some parties due to the incorporation of chemical or biological residues obtained after the treatment of wastewaters even after the stabilization process [3,4,79], leading to the acceptance or rejection for the incorporation in agriculture in agreement with the regulatory policies. In the case of the WWTP of the current study, it is not possible to use the by-products after the stabilization process, because it exceeds the maximum limit of microorganisms allowed [7,21]. Stabilization Time (ST) was identified as an environmental variable, that even if it did not have a high dependence level for the system, it could be considered, as it might affect other variables involved in the reutilization process.

4.2. Evaluation of Stakeholders in the Management and Use of Biosolids, Mactor

The influence and dependence positions of the determining actors for the management and use of biosolids (Figure 4) mainly showed the impact of institutional and control parties which were identified as high influence/low dependence actors highlighting their importance for the system. This makes them key players in the system as are involved in the strategic planning and regulations for the field, approving or revoking processes among the WWTPs [80]. Additionally, it is important to keep in mind considerations of the community and the people as their participations are relevant in terms of establishing objectives, interests, concerns, or restrictions on the use of the biosolids and by-products and could exert some pressure on the field (Figure 4B).

Liaison actors identified for both management and use of biosolids (Figure 4A,B) require adequate coordination and agreements between the parties, since they are prone to generate conflicts among themselves, giving, as a result, isolated and discontinuous participation [81]. Moreover, internal variables (Figure 4), including Community, Mayoralty, and WWTP operators for the WWTP management and academia for biosolids use, are conceived as dependent actors due to the subordinated conditions that some may have, depending on the actions, projects, and decision-making from the stakeholders of the liaison. Biosolids management had led to the identification and achievement of goals and opportunities for improvement based on the interaction of different stakeholders which had some levels of resilience to the application of these by-products as reported in previous studies [75,82].

As for the six challenges proposed in Table 4, disagreements regarding the development and fulfillment of those were concentrated on three main actors of the WWTP: WWTP-Mang, WWTP-FA, and WTTP-OTD. This opposition may be due to the additional costs that might need to be assumed for the proper and periodic maintenance of the infrastructure, the implementation of new systems, controls, and the certification of the by-products as organic. Thus, besides the main budget line for the sludges and biosolids treatment, entities should contemplate in their financial statements a budget line for contingencies, which increases the total budget, forcing them to seek and manage resources for both long- and short-term sustainability [60,61,72]. Another important factor for the economic impact of the WWTP is that the Water and Sewer Utility bills in Colombia do not include the total costs for adequate management of the WWTP, including the treatment and stabilization of biosolids, apart from the administrative costs [83]. Keeping updated the information will favor the sustainability of the WWTPs along the country based on the needs of each region and its management of wastewaters.

4.3. Social Network Analysis (SNA) for the Management and Use of Biosolids in Agriculture

SNA analysis for the current study showed a low-density level (0.28) for the management of the WWTP and use of the biosolids, involving few actors in the degrees of centrality and betweenness levels based on the communication and interaction levels between WWTPs located in different cities/municipalities of Colombia and other stakeholders (Figure 6). These low-density values are the result of weak interactions among stakeholders in the systems leading to fragmentation issues, such as reduced feedback and low cooperation between the different actors, which could impact the operation of the WWTPs and the utilization of biosolids [80]. With these results, we identified one of the main concerns as the articulation and functionality of the WWTPs and their impact on the community. Low interactions and weak communication flux between parties result in the overall perception of the community as well as for other external entities. Currently, there is not a good perception of the system by the community, as some people describe it as having a poor (25.9%) or intermediate (37%) performance [23].

With the results obtained in this study, we identify good opportunities for improvement as the implementation of a new system in the future involving more novel actors could favor a better operation, management, and further applications of the WWTPs and their by-products. Figures 7 and 8 suggest that if working in collaboration with other stakeholders, it would be possible to improve the communication among them and to enhance networks among stakeholders obtaining density levels of about 0.49 or 0.57, and a higher closeness centrality and betweenness centrality. Giving, as a result, a direct interaction between parties for improving the performance of the system [84], facilitating access to the information, and acting as mediators for internal and external parties, even in the case of decision-making.

Implementing other actors for the management of the organizations favors the decentralization of the system leading to better organizational and planning opportunities and strengthening the networks at horizontal and vertical levels, causing significant changes in the systems. Despite this, there will still be other challenges to the articulation of the systems that will have to be mediated by key actors, intermediaries, and information disseminators [45,49,85]. The development of a network with better connections, density, and centrality values will be favorable for the system, since currently, most public and private entities have shown a willingness to participate in the improvement of biosolids management and quality, as well as a favorable position (>64%) regarding the use of biosolids in agriculture [23].

4.4. Management Framework and Methods Analysis (MicMac, Mactor, and SNA)

MicMac and Mactor analyses of the system based on the influence and dependence of variables and actors are good and useful prospective strategies, as those helped us to identify the determining variables and actors that have a real impact on the management of the WWTP and further use of biosolids. Key actors and variables would be the basis for the system to achieve and fulfill the proposed challenges and objectives, leading to an improvement and significant change in the functionality of the WWTP. However, one of the limitations of the Mactor strategy is that it only considers the incidence level of stakeholders in terms of existence, mission, projects, and processes that could be involved, leaving aside other aspects, such as communication and direct work. Levels of communication and direct execution that could carry out specific actors in the system, despite their power and influence, become important characteristics due to the operational capacities they could represent, changing the roles of key actors in terms of information dissemination and networking with other communities and stakeholders. Therefore, the consideration of other method analyses, such as the SNA strategy, is essential for a better understanding of the role and interactions of stakeholders and variables for the evaluated systems.

The integrated analysis used in this study allowed us to identify the main factors related to the management of the WWTP and treatment of biosolids, related to influence and dependence of stakeholders as well as the identification of opportunities for improvement in the communication and workflow for the fulfillment of the objectives. Our results and the applications of the merged analysis of MicMac/Mactor and SNA will complement the traditional management of the WWTP in terms of (I) pathogens reduction, (II) further applications of biosolids, (III) operational skills, (IV) organic fertilizers market studies, and (V) possible applications of by-products to soils with agricultural and veterinarian approaches [23,86].

Our results are an essential contribution to the development of CONPES 4004 [32], aimed at identifying the potential use of wastewater by-products and integrating them under the circular economy strategy, as well as new business opportunities. This is the first study in Colombia in which prospective strategies are used and integrated with SNA in terms of management of sludges and biosolids with further applications in agriculture. Structural analysis, stakeholders' evaluation, and social network analysis allowed us to understand the current execution of the system, but also to identify future scenarios that may improve WWTP management and biosolids re-utilization, giving the basis for planning new challenges and breaking barriers for integration/articulation of interested parties and decision-making entities.

According to EU data, the use of dried sludge for agriculture has been increasing in recent years and is projected to keep growing in the coming years [13,16,87,88]. This effect is largely due to the existing legislation on the management of this waste and the formulation of a new regulation focused on the inclusion of solid waste in the circular economy model that would favor their production, recycling, and disposal on land suitable for agriculture [13]. Additionally, in recent years the European Union has adopted an ambitious circular economy plan that promotes the reuse, recycling, and recovery of waste, allowing the classic vision of WWTPs to be changed, projecting it towards a more sustainable model [28,29,89]. On the other hand, the approach taken by the European community is interdisciplinary, considering the economic and environmental complexities of the sites where waste management mechanisms are developed [90].

The opposite is the case in developing countries in which the management and use of solid waste (biosolids) are not one of the main objectives, even though these present high biological risks in various scenarios. Equally, this situation entails the formulation of challenges due to the complexity of this process in making decisions to develop an efficient, effective, and sustainable process [36,91–93]. Within the different improvement processes in waste management, the following have been proposed: coherent and effective governance models, the acceleration of the transition to a circular economy, the promotion of participation of local scenarios, the generation of data and information to understand and improve waste management, and the coordination of objectives between national, municipal and local entities [94–97].

The execution and analysis of future scenarios and the analysis of networks from the vision of a municipality in Colombia becomes a key that allows the relationship and prioritization of various variables, challenges or objectives, validation of the positions, and inclusion of new actors or decision-makers under the approach of improving the internal management of biosolids in a WWTP, as well as the subsequent use of these allowing the creation of bases for organization, participation, and recognition of new actors.

The analysis of this study corroborates that, although progress has been made in Latin America and specifically in Colombia, in the implementation of policies and regulations regarding the management, disposal, and reuse of biosolids, an organizational change must be achieved in which the actors involved understand the importance of adequate reuse of this waste to take advantage of it inland destined for agriculture and thus incorporate it into the productive cycle under the circular economy model.

The legislation in the country advanced in this specific aspect with the issuance of decree 1287 of 2014 [21] by which criteria are established for the use of biosolids generated in municipal wastewater treatment plants. However, this is not fully met due to the existence of technical, structural, financial, cultural, and social difficulties in the WWTPs and their areas and population of influence. On the other hand, currently in Colombia, there has been progress in the regulation of the reuse of wastewater (Resolution 1256 of 2021) [98]; with the implementation of this, a general cultural change is expected that will lead to the acceptance of these residues by the community, including both residual water and biosolids. Therefore, it is inferred that future studies and actions should be aimed at generating changes in the bases of the biosolids reuse model and system.

Nowadays, Colombia's economic and social policy CONPES 4004 [32] promotes the use of biosolids through the (I) analysis of biosolids as a potential element in the production cycle and (II) definition of business opportunities, which allows us to have future experiences of use, similar to countries such as Mexico, Chile, Argentina and Paraná (Brazil). Paraná is one of the main cities where biosolids tend to be used in agriculture as a priority disposal, thus favoring the agricultural potential of the area [13,99,100].

On the other hand, regulations in Latin America are mainly adaptations of regulations from industrialized countries. This type of the adoption of standards would not allow them to be adjusted to the needs and/or conditions of each region [101], so complementary guidelines have been presented to decree 1287 of 2014 [21], in which technical and methodological aspects are proposed, such as improving the storage, transport of biosolids, and application in degraded, agricultural, and forestry soils, allowing us to improve good management practices and reuse [102].

Although this research does not contemplate or evaluate Decision Support Systems (DSS), these are systems or methods that gain relevance for the information they can contribute to the solution of unstructured problems [103,104]. All this is from available models and data, which have improved the efficiency of decision-making in strategic issues, such as water and sewage sludge management [34,104–109], through the involvement of multiple variables, as well as the consideration of technical, regulatory, socio-environmental, economic, and administrative aspects. However, the availability and provision of data, as well as the interpretation, must be done from a holistic perspective, thus becoming fundamental agents for the correct formulation, optimization, interpretation, and application of the DSS [34,107], creating in the end a synergy between the different chains evaluated and analyzed.

5. Conclusions

Prospective analysis of the WWTP management and use of biosolids identified key variables and actors that require immediate action, especially with an integrated and articulated strategy for the improvement of the overall system.

Stakeholders within the WWTP will require special attention for the formulation, compliance, monitoring, and control of the objectives, mainly for the implementation of the institution's environmental policy (IEP) for sludge and biosolids, the improvement in infrastructure (WWTP- Infr_), and treatment efficiency (TE). On the other hand, at the agricultural level, the following are required: the certification of products (Cert_), the definitions of the classes of biosolids to be generated (BT) and their uses, and the establishment of the distribution mechanism for the waste as well as the benefits granted by management and use (BM).

As a low connectivity between stakeholders that are present today was identified in the SNA analysis, it is suggested to develop new connections encouraging the participation of novel stakeholders which could favor the system, improving the communication skills and collaborative work.

Our results provide new insights into the traditional management of sludges and biosolids in Colombia, which will help in the improvement of the WWTPs in the country and further applications of the biosolids in the agriculture field and some other scenarios.

Likewise, it makes visible the need for greater coordination and interaction between both traditional and new actors, which would improve the density of current networks for the management and use of biosolids, and strengthen the studies and evaluations in the future of the actions and the presence of a greater number of decision-making agents and stakeholders.

Author Contributions: C.V., A.C.S.-A. and F.-J.V. participated in the search and collection of information. C.V. consolidated, tabulated, analyzed the data, and drafted the manuscript. A.C.S.-A., F.-J.V., A.M., C.C.-Z. and M.G.M. reviewed, adjusted, and co-edited the document. C.C.-Z. attained financial support. C.V. and M.G.M. carried out the conception of the study. M.G.M. conducted on, the direction of the study. All authors have read and agreed to the published version of the manuscript.

Funding: This research was funded by the Pontificia Universidad Javeriana-Bogotá, Colombia, grant number 005874, and the publication by Ministerio de Ciencia, Tecnología e Innovación—MINCIENCIAS, grant number 922-2022.

Institutional Review Board Statement: Not applicable.

Informed Consent Statement: Not applicable.

Data Availability Statement: Not applicable.

Acknowledgments: The authors thank the different public and private entities and each of the people who participated in the surveys and interviews conducted during the period 2020–2021. They are also grateful for the information and documentation provided by the WWTP of the study site and other entities. The authors would like to thank Ana Lorena Rojas Sabogal from the Specialized Services of the library Pontificia Universidad Javeriana for her collaboration and support in the management and interpretation of the results obtained with the Gephi software. The authors would like to thank Nury Olaya (nury.olaya@urosario.edu.co) for the translation and English proofreading of the manuscript. We thank the engineer Germán Florez Barrera for the elaboration of the map.

Conflicts of Interest: The authors declare no conflict of interest.

Appendix A

Table A1. Roles of institutions and stakeholders involved in the management and use of biosolids for agriculture.

Stakeholders (n = 53)	Operating Level	Associated Functions
Acad_ (n = 7)	CM	Integrate research, academia, and social projection from teaching, education, and service.
AI (n = 9)	N	To provide economic income and support to the farmers. Reduce post-harvest losses in agricultural production. Develop new forms of production
CAR (n = 1)	N	"Maximum environmental authority under the criteria and guidelines established by the Ministry of Environment and sustainable development." "Promote and develop community participation in activities and programs for environmental protection, sustainable development, and adequate management of renewable natural resources."
DEyA (n = 1)	M	"Define programs for entrepreneurship and agricultural development, providing technical assistance to all the agents involved, adopting and directing the plans that the municipality needs to advance for the development of this sector, especially the farming sector." "Promote community participation and the social improvement of the agricultural activity of the residents of the municipality, taking into account the mechanisms of citizen participation and the needs of the community."

Table A1. *Cont.*

Stakeholders (n = 53)	Operating Level	Associated Functions
DNP (n = 1)	N	"To design, guide and evaluate Colombian public policies, the management, and allocation of public investment and the implementation of these in plans, programs and projects of the government in the social, economic and environmental fields."
ESPB	D	"Manage the provision and strengthening of public services in the department of Boyacá, providing support, advice, and technical assistance at the municipal and regional levels."
F/A (n = 7)	N	Maintain agricultural activities and the development of the national and local economy.
WWTP-FA	D	"Optimize the company's own and financial resources to guarantee the fulfillment of its objectives."
ICA (n = 1)	N	"Exercise technical control over the production, importation, and commercialization of agricultural inputs to prevent risks that may affect agricultural health."
MinA_	N	"To design and regulate public policies and general conditions for environmental sanitation.... to prevent, repress, eliminate or mitigate the impact of polluting, deteriorating or destructive activities on the environment or natural heritage, in all economic and productive sectors."
MinV_	N	"Define feasibility and eligibility criteria for water, sewerage and sanitation projects and approve them, and provide technical assistance to territorial entities, environmental authorities, and public utility service providers."
Ope_	D	"Carry out all the necessary activities so that the wastewater treatment plant remains in good condition"
SSPD	N	"To monitor, inspect and control compliance by the supervised parties with the provisions that regulate the proper rendering of residential public utilities and the protection of users."
WWTP-OTD	D	"Project, carry out, and supervise the infrastructure works necessary for the efficient provision of services."
WM_ (n = 11)	N	Collect organic waste to be treated or disposed of correctly.
WWTP-EM (n = 3)	M	"Establish and implement actions aimed at directing the environmental management of the company operating the WWTP; ensure compliance with environmental standards; promote cleaner production practices and the rational use of natural resources."
WWTPs (n = 12)	CM	"Guarantee to the community the treatment of wastewater in the coverage area to reduce the environmental impact, through the correct operation of the WWTP and maintenance of its components."

CM: cities and Municipalities, D: departmental, and N: national.

References

1. Paes, L.A.B.; Bezerra, B.S.; Deus, R.M.; Jugend, D.; Battistelle, R.A.G. Organic Solid Waste Management in A Circular Economy Perspective–A Systematic Review and SWOT Analysis. *J. Clean. Prod.* **2019**, *239*, 118086. [CrossRef]
2. Srivastava, P.K.; Kulshreshtha, K.; Mohanty, C.S.; Pushpangadan, P.; Singh, A. Stakeholder-Based SWOT Analysis for Successful Municipal Solid Waste Management in Lucknow, India. *J. Waste Manag.* **2005**, *25*, 531–537. [CrossRef] [PubMed]
3. Wijesekara, H.; Bolan, N.S.; Kumarathilaka, P.; Geekiyanage, N.; Kunhikrishnan, A.; Seshadri, B.; Saint, C.; Surapaneni, A.; Vithanage, M. Biosolids Enhance Mine Site Rehabilitation and Revegetation. In *Environmental Materials and Waste*; Prasad, M.N.V., Shih, K., Eds.; Academic Press: Cambridge, MA, USA, 2016; pp. 45–71. ISBN 9780128038376.
4. Basic Information about Biosolids. Available online: https://www.epa.gov/biosolids/basic-information-about-biosolids (accessed on 24 February 2021).
5. Shammas, N.K.; Wang, L.K. Transport and Pumping of Sewage Sludge and Biosolids. In *Biosolids Engineering and Management*; Humana Press: Totowa, NJ, USA, 2008; pp. 1–64.

6. Zhang, W.; Alvarez-Gaitan, J.; Dastyar, W.; Saint, C.; Zhao, M.; Short, M. Value-Added Products Derived from Waste Activated Sludge: A Biorefinery Perspective. *Water* **2018**, *10*, 545. [CrossRef]
7. Venegas, C.; Sánchez-Alfonso, A.C.; Celis Zambrano, C.; González Mendez, M.; Vesga, F.-J. *E. coli* CB390 as an Indicator of Total Coliphages for Microbiological Assessment of Lime and Drying Bed Treated Sludge. *Water* **2021**, *13*, 1833. [CrossRef]
8. MinVivienda. *Plan Director de Agua Y Saneamiento Básico 2018–2030*; MinVivienda: Bogotá, Colombia, 2018.
9. Porcentaje de Aguas Residuales Urbanas Domésticas Tratadas de Manera Segura-Indicador 6.3.1.P. Available online: https://www.ods.gov.co/es/data-explorer?state=%7B%22goal%22%3A%226%22%2C%22indicator%22%3A%226.3.1.P%22%2C%22dimension%22%3A%22COUNTRY%22%2C%22view%22%3A%22line%22%7D (accessed on 17 April 2022).
10. Agua Limpia Y Saneamiento-La Agenda 2030 en Colombia-Objetivos de Desarrollo Sostenible. Available online: https://www.ods.gov.co/es/objetivos/agua-limpia-y-saneamiento (accessed on 14 January 2021).
11. DNP. *Estrategia Para la Implementación de los Objetivos de Desarrollo Sostenible (ODS) en Colombia, CONPES 3918*; Departamento Nacional de Planeación (DNP): Bogotá, Colombia, 2018.
12. SSPD. Estudio Sectorial de los Servicios Públicos Domiciliarios de Acueducto Y Alcantarillado 2014–2017. In *Superintendencia de Servicios Públicos Domiciliarios*; Superintendencia de Servicios Públicos Domiciliarios: Bogotá, Colombia, 2018; pp. 1–88.
13. Wiśniowska, E.; Grobelak, A.; Kokot, P.; Kacprzak, M. Sludge Legislation-Comparison between Different Countries. In *Industrial and Municipal Sludge: Emerging Concerns and Scope for Resource Recovery*; Prasad, M.N.V., de Campos Favas, P.J., Vithanage, M., Mohan, S.V., Eds.; Elsevier: Amsterdam, The Netherlands, 2019; pp. 201–224. ISBN 9780128159071.
14. SSPD. Estudio Sectorial de los Servicios Públicos Domiciliarios de Acueducto y Alcantarillado-2019. In *Superintendencia de Servicios Públicos Domiciliarios*; Superintendencia de Servicios Públicos Domiciliarios: Bogotá, Colombia, 2020; pp. 1–64.
15. Spinosa, L. *Wastewater Sludge: A Global Overview of the Current Status and Future Prospects*, 2nd ed.; IWA Publishing: Kota Banjarmasin, Indonesia, 2011; Volume 10, ISBN 9781780401195.
16. Sewage Sludge Production and Disposal from Urban Wastewater. Available online: https://ec.europa.eu/eurostat/web/environment/water (accessed on 14 November 2020).
17. Martin-Hurtado, R.; Nolasco, D. *Managing Wastewater as a Resource in Latin America and the Caribbean Towards a Circular Economy Approach*; World Bank: Washington, DC, USA, 2017; pp. 1–28.
18. Aguas Andinas. *Reporte de Sostenibilidad Pura Vida 2013*; Aguas Andinas: Santiago, Chile, 2014; pp. 1–132.
19. Rosales, E.P. Aplicación Benéfica de Biosólidos en Chile: Desafíos, Dificultades y Oportunidades de Mejora. *Revista AIDIS* **2018**, *57*, 18–23. Available online: https://www.aidis.cl/wp-content/uploads/2016/10/REVISTA-AIDIS-AGOSTO-2018-final.pdf (accessed on 6 February 2021).
20. Bittencourt, S. Agricultural Use of Sewage Sludge in Paraná State, Brazil: A Decade of National Regulation. *Recycling* **2018**, *3*, 53. [CrossRef]
21. MinVivienda. *Se Establecen Criterios Para El Uso de los Biosólidos Generados en Plantas de Tratamiento de Aguas Residuales en El Territorio de Colombia, Decreto 1287*; Ministerio de Vivienda, Ciudad y Territorio: Bogotá, Colombia, 2014; pp. 1–15.
22. ICONTEC. *Productos Para La Industria Agrícola. Productos Orgánicos Usados Como Abonos O Fertilizantes Y Enmiendas O Acondicionadores de Suelo, NTC 5167*; Instituto Colombiano de Normas Técnicas y Certificación (ICONTEC): Bogotá, Colombia, 2011; pp. 1–51.
23. Venegas, C.; Sánchez-Alfonso, A.C.; Celis, C.; Vesga, F.-J.; Mendez, M.G. Management Strategies and Stakeholders Analysis to Strengthen the Management and Use of Biosolids in a Colombian Municipality. *Sustainability* **2021**, *13*, 12180. [CrossRef]
24. EMPAS. *Informe de Gestión 2019*; Empresa Pública de Alcantarillado de Santander S.A. E.S.P: Santander, Colombia, 2019; pp. 1–169.
25. EPM Air and Soil Care, an EPM Contribution for the Health of the Earth, Empresas Públicas de Medellín-EPM. Available online: https://www.epm.com.co/site/cuidado-del-aire-y-de-los-suelos-un-aporte-de-epm-por-la-salud-de-la-tierra (accessed on 8 January 2021).
26. Montoya, G.G.; Gómez, C.X.R. *Acondicionadores de Suelo y Fertilizantes a Partir de Biosólidos Generados en Plantas de Tratamiento de Aguas Residuales de EPM*; Empresas Públicas de Medellín: Medellín, Colombia, 2019; pp. 11–21.
27. González, R.; Guzmán, G.; Mayorga, C. Avances en El Saneamiento y la Gestión de Biosólidos en Colombia. *Rev. FACCEA Univ. Amazon.* **2019**, *9*, 113–126. [CrossRef]
28. Cornejo, P.K.; Becker, J.; Pagilla, K.; Mo, W.; Zhang, Q.; Mihelcic, J.R.; Chandran, K.; Sturm, B.; Yeh, D.; Rosso, D. Sustainability Metrics for Assessing Water Resource Recovery Facilities of the Future. *Water Environ. Res.* **2019**, *91*, 45–53. [CrossRef]
29. Peng, S.; Cui, H.; Ji, M. Research on the Sustainable Water Recycling System at Tianjin University's New Campus. In *Unmaking Waste in Production and Consumption: Towards the Circular Economy*; Crocker, R., Saint, C., Chen, G., Tong, Y., Eds.; Emerald Publishing Limited: Bingley, UK, 2018; pp. 295–307.
30. Gherghel, A.; Teodosiu, C.; de Gisi, S. A Review on Wastewater Sludge Valorisation and Its Challenges in the context of Circular Economy. *J. Clean. Prod.* **2019**, *228*, 244–263. [CrossRef]
31. Neczaj, E.; Grosser, A. Circular Economy in Wastewater Treatment Plant–Challenges and Barriers. *Proc. West. Mark Ed. Assoc. Conf.* **2018**, *2*, 614. [CrossRef]
32. DNP. *Economía Circular en La Gestión de los Servicios de Agua Potable Y Manejo de Aguas Residuales, CONPES 4004*; Departamento Nacional de Planeación: Bogotá, Colombia, 2020; pp. 1–64.

33. Axelrad, G.; Gershfeld, T.; Feinerman, E. Reclamation of Sewage Sludge for Use in Israeli Agriculture: Economic, Environmental and Organizational Aspects. *J. Environ. Plan. Manag.* **2013**, *56*, 1419–1448. [CrossRef]
34. Bertanza, G.; Baroni, P.; Canato, M. Ranking Sewage Sludge Management Strategies by Means of Decision Support Systems: A Case Study. *Resour. Conserv. Recycl.* **2016**, *110*, 1–15. [CrossRef]
35. Laura, F.; Tamara, A.; Müller, A.; Hiroshan, H.; Christina, D.; Serena, C. Selecting Sustainable Sewage Sludge Reuse Options through a Systematic Assessment Framework: Methodology and Case Study in Latin America. *J. Clean. Prod.* **2020**, *242*, 1–12. [CrossRef]
36. Eggers, S.; Thorne, S. Conducting Effective Outreach with Community Stakeholders About Biosolids: A Customized Strategic Risk Communications Process™ Based on Mental Modeling. In *Mental Modeling Approach*; Springer: New York, NY, USA, 2017; pp. 153–177.
37. Staff Relocation Services Colombia (SRS) Colombia: Global Leader in Agriculture. Available online: http://www.relocationsrs.com.co/colombia-world-agricultural-pantry/ (accessed on 21 February 2021).
38. Strategic Foresight La Prospective Problems and Methods. Available online: http://www.laprospective.fr/dyn/francais/memoire/strategicforesight.pdf (accessed on 16 March 2021).
39. Godet, M. *De La Anticipación A La Acción: Manual de Prospectiva Y Estrategia*; Marcombo Boixareu Editore, Ed.; Barcelona, Spain, 1993; Volume 5, pp. 1–380. Available online: https://administracion.uexternado.edu.co/matdi/clap/De%20la%20anticipaci%C3%B3n%20a%20la%20acci%C3%B3n.pdf (accessed on 18 October 2021).
40. Caniato, M.; Vaccari, M.; Visvanathan, C.; Zurbrügg, C. Using Social Network and Stakeholder Analysis to Help Evaluate Infectious Waste Management: A Step towards a Holistic Assessment. *J. Waste Manag.* **2014**, *34*, 938–951. [CrossRef]
41. Tools for Knowledge and Learning: A Guide for Development and Humanitarian Organisations. Available online: https://cdn.odi.org/media/documents/188.pdf (accessed on 8 September 2021).
42. Holland, J. *Tools for Institutional, Political, and Social Analysis of Policy Reform: A Sourcebook for Development Practitioners*; The World Bank: Washington, DC, USA, 2007; ISBN 978-0-8213-6890-9.
43. Reed, M.S.; Graves, A.; Dandy, N.; Posthumus, H.; Hubacek, K.; Morris, J.; Prell, C.; Quinn, C.H.; Stringer, L.C. Who's in and Why? A Typology of Stakeholder Analysis Methods for Natural Resource Management. *J. Environ. Manag.* **2009**, *90*, 1933–1949. [CrossRef]
44. Vance-Borland, K.; Holley, J. Conservation Stakeholder Network Mapping, Analysis, and Weaving. *Conserv. Lett.* **2011**, *4*, 278–288. [CrossRef]
45. Putra, D.I.; Matsuyuki, M. Disaster Management Following Decentralization in Indonesia: Regulation, Institutional Establishment, Planning, and Budgeting. *J. Disaster Res.* **2019**, *14*, 173–187. [CrossRef]
46. Scott, J.; Carrington, P. *The SAGE Handbook of Social Network Analysis*; SAGE Publications Ltd.: London, UK, 2014; ISBN 9781847873958.
47. Methods of Prospective. Available online: http://en.laprospective.fr/methods-of-prospective.html (accessed on 28 February 2021).
48. Bastian, M.; Heymann, S.; Jacomy, M. Gephi: An Open Source Software for Exploring and Manipulating Networks. In Proceedings of the The Third International AAAI Conference on Weblogs and Social Media (ICWSM-09), San Jose, CA, USA, 17–20 May 2009; pp. 361–362.
49. Prell, C.; Hubacek, K.; Reed, M. Stakeholder Analysis and Social Network Analysis in Natural Resource Management. *Soc. Nat. Resour.* **2009**, *22*, 501–518. [CrossRef]
50. PTAR. Planta de Tratamiento de Aguas Residuales Domésticas, Sitio de Estudio. In *Informe Semestral de Operaciones PTAR 2019–2020*; PTAR: Boyacá, Colombia, 2021; pp. 1–80.
51. Langergraber, G.; Pressl, A.; Kretschmer, F.; Weissenbacher, N. Small Wastewater Treatment Plants in Austria–Technologies, Management and Training of Operators. *Ecol. Eng.* **2018**, *120*, 164–169. [CrossRef]
52. Murillo Gómez, D.F.; Hernández Garzón, C.A.; Torres Córdoba, D.M. Gobernanza Del Agua en Colombia Como Política Pública Y de Responsabilidad Social en Villavicencio: Estudio de Casos. Bachelor's Thesis, Universidad Cooperativa de Villavicencio, Villavicencio, Colombia, 2019.
53. Kashmanian, R.M.; Kluchinski, D.; Richard, T.L.; Walker, J.M. Quantities, Characteristics, Barriers, and Incentives for Use of Organic Municipal By-Products. In *Land Application of Agricultural, Industrial, and Municipal By-Products*; Power, J.F., Dick, W.A., Kashmanian, R.M., Sims, J.T., Wright, R.J., Dawson, M.D., Bezdicek, D., Eds.; John Wiley & Sons, Inc.: Hoboken, NJ, USA, 2018; pp. 127–167.
54. EPA. *A Plain English Guide to the EPA Part 503 Biosolids Rule*; Environmental Protection Agency (EPA): Washington, DC, USA, 1994; pp. 1–183.
55. World Bank Group; GSWP; PPIAF. *From Waste to Resource-Shifting Paradigms for Smarter Wastewater Interventions in Latin America and the Caribbean: Background Paper IV. Policy, Regulatory, and Institutional Incentives for the Development of Resource Recovery Projects in Wastewater*; World Bank Group, Global Water Security & Sanitation Partnership (GWSP), PPIAF, Eds.; World Bank Publications: Washington, DC, USA, 2019; pp. 1–26.
56. Otoo, M.; Drechsel, P.; Hanjra, M.A. Business Models and Economic Approaches for Nutrient Recovery from Wastewater and Fecal Sludge. In *Wastewater*; Springer: Dordrecht, The Netherlands, 2015; pp. 247–268.
57. Collivignarelli, M.C.; Abbà, A.; Carnevale Miino, M.; Torretta, V. What Advanced Treatments Can Be Used to Minimize the Production of Sewage Sludge in WWTPs? *Appl. Sci.* **2019**, *9*, 650. [CrossRef]

58. Rulkens, W.H. Sustainable Sludge Management-What Are the Challenges for the Future? *Water Sci. Technol.* **2004**, *49*, 11–19. [CrossRef] [PubMed]
59. Tytła, M. Identification of the Chemical Forms of Heavy Metals in Municipal Sewage Sludge as a Critical Element of Ecological Risk Assessment in Terms of Its Agricultural or Natural Use. *Int. J. Environ. Res. Public Health* **2020**, *17*, 4640. [CrossRef] [PubMed]
60. Andreoli, C.V.; von Sperling, M.; Fernandes, F. Sludge Treatment and Disposal. *Water Intell. Online* **2007**, *6*, 1–258. [CrossRef]
61. Gray, N.F. Sludge Treatment and Disposal. In *Water Science and Technology: An Introduction for Environmental Scientists and Engineers*; Elsevier Ltd.: Amsterdam, The Netherlands, 2010; pp. 645–685. ISBN 978-1-85617-705-4.
62. Vivienda. *Reglamento Para El Reaprovechamiento de los Lodos Generados en Las Plantas de Tratamiento de Aguas Residuales. Decreto Supremo N° 015-2017*; Ministerio de Vivienda, Construcción y Saneamiento: Lima, Perú, 2017; pp. 1–9.
63. Vivienda. *Protocolo de Monitoreo de Biosólidos-Resolución Ministerial N° 093-2018*; Ministerio de Vivienda, Construcción y Saneamiento: Lima, Perú, 2018; pp. 1–72.
64. MAyDS. *Norma Técnica Para El Manejo Sustentable de Barros Y Biosólidos Generados en Plantas Depuradoras de Efluentes Líquidos Cloacales Y Mixtos Cloacales-Industriales en Argentina, Resolución 410/2018*; Ministerio de Ambiente Y Desarrollo Sostenible: Ciudad Autónoma de Buenos Aires, Argentina, 2018; pp. 1–15.
65. Conama. *Define Critérios E Procedimentos, Para O Uso Agrícola de Lodos de Esgoto Gerados Em Estações de Tratamento de Esgoto Sanitário E Seus Produtos Derivados, E Dá Outras Providências, Resolução N° 375*; Conselho Nacional do Meio Ambiente: Brasília, Brasil, 2006; pp. 1–32.
66. Amorim Junior, S.S.; Hwa Mazucato, V.S.; Machado, B.d.S.; de Oliveira Guilherme, D.; Brito da Costa, R.; Correa Magalhães Filho, F.J. Agronomic Potential of Biosolids for a Sustainable Sanitation Management in Brazil: Nutrient Recycling, Pathogens and Micropollutants. *J. Clean. Prod.* **2021**, *289*, 1–9. [CrossRef]
67. Díaz Gómez, J.; Cifuentes Osorio, G.R. *De La Generación Al Aprovechamiento Sostenible: Lodos Y Biosólidos de Tratamiento de Aguas Y Aguas Residuales*, 1st ed.; Díaz Gómez, J., Cifuentes Osorio, G.R., Eds.; Universidad de Boyacá: Boyacá, Colombia, 2020; ISBN 9789585120136.
68. ECLAC. *Access to Information, Participation and Justice in Environmental Matters in Latin America and the Caribbean: Towards Achievement of the 2030 Agenda for Sustainable Development (LC/TS.2017/83)*; United Nations Publication: Santiago, Chile, 2018; pp. 1–146.
69. Ensure Transparency and Access to Information. Available online: https://socialprotection-humanrights.org/framework/principles/ensure-transparency-and-access-to-information/ (accessed on 27 October 2021).
70. Moya, B.; Sakrabani, R.; Parker, A. Realizing the Circular Economy for Sanitation: Assessing Enabling Conditions and Barriers to the Commercialization of Human Excreta Derived Fertilizer in Haiti and Kenya. *Sustainability* **2019**, *11*, 3154. [CrossRef]
71. Murray, A.; Mekala, G.D.; Chen, X. Evolving Policies and the Roles of Public and Private Stakeholders in Wastewater and Faecal-Sludge Management in India, China and Ghana. *Water Inter.* **2011**, *36*, 491–504. [CrossRef]
72. Gale, A.J. The Australasian Biosolids Partnership and Public Perceptions. *Water Pract. Technol.* **2007**, *2*. [CrossRef]
73. Danso, G.; Otoo, M.; Ekere, W.; Ddungu, S.; Madurangi, G. Market Feasibility of Faecal Sludge and Municipal Solid Waste-Based Compost as Measured by Farmers' Willingness-to-Pay for Product Attributes: Evidence from Kampala, Uganda. *Resources* **2017**, *6*, 31. [CrossRef]
74. Guerrero, L.A.; Maas, G.; Hogland, W. Solid Waste Management Challenges for Cities in Developing Countries. *Waste Manag.* **2013**, *33*, 220–232. [CrossRef]
75. Beecher, N.; Connell, B.; Epstein, E.; Filtz, J.; Goldstein, N.; Lono, M. Public Perception of Biosolids Recycling: Developing Public Participation and Earning Trust. *Water Intell. Online* **2004**, *3*, 1–140. [CrossRef]
76. Lindsay, B.E.; Zhou, H.; Halstead, J.M. Factors Influencing Resident Attitudes Regarding the Land Application of Biosolids. *Am. J. Altern. Agric.* **2000**, *15*, 88–95. [CrossRef]
77. Krogmann, U.; Gibson, V.; Chess, C. Land Application of Sewage Sludge: Perceptions of New Jersey Vegetable Farmers. *WM&R* **2001**, *19*, 115–125. [CrossRef]
78. Whitehouse, S.; Tsigaris, P.; Wood, J.; Fraser, L.H. Biosolids in Western Canada: A Case Study on Public Risk Perception and Factors Influencing Public Attitudes. *Environ. Manag.* **2021**, *69*, 179–195. [CrossRef]
79. Margot, J.; Rossi, L.; Barry, D.A.; Holliger, C. A Review of the Fate of Micropollutants in Wastewater Treatment Plants. *WIREs Water.* **2015**, 457–487. [CrossRef]
80. Le, N.; Nguyen, T.; Zhu, D. Understanding the Stakeholders' Involvement in Utilizing Municipal Solid Waste in Agriculture through Composting: A Case Study of Hanoi, Vietnam. *Sustainability* **2018**, *10*, 2314. [CrossRef]
81. Garza Villegas, J.B.; Cortez Alejandro, D.V. El Uso Del Método MICMAC y MACTOR Análisis Prospectivo en Un Área Operativa Para La Búsqueda de La Excelencia Operativa a Través Del Lean Manufacturing. *Innov. Negocios* **2017**, *8*, 335–356. [CrossRef]
82. Pepper, I.L.; Zerzghi, H.; Brooks, J.P.; Gerba, C.P. Sustainability of Land Application of Class B Biosolids. *J. Environ. Qual.* **2008**, *37*, S-58–S-67. [CrossRef]
83. CRA. *Diagnóstico E Identificación de Problema, Objetivos Y Alternativas Y Regulación Frente A Tarifas Por Actividad Del Servicio Tratamiento de Vertimientos*; Comisión de Regulación de Agua Potable y Sanemaiento Basico (CRA): Bogotá, Colombia, 2019; pp. 1–115.
84. Lienert, J.; Schnetzer, F.; Ingold, K. Stakeholder Analysis Combined with Social Network Analysis Provides Fine-Grained Insights into Water Infrastructure Planning Processes. *J. Environ. Manag.* **2013**, *125*, 134–148. [CrossRef]

85. OECD. *Sustainable Development Strategies: A Resource Book*, 1st ed.; Bass, S., Dalal-Clayton, B., Eds.; Routledge: London, UK, 2012; ISBN 9781849772761.
86. Camilo, V. Aprovechamiento de los Biosólidos Para La Agricultura A Través Del Fortalecimiento de Estrategias de Gestión Ambiental Para Un Municipio de Boyacá. Master's Thesis, Pontificia Universidad Javeriana, Bogotá, Colombia, 2021.
87. Part III Sustainable and Safe Re-Use of Municipal Sewage Sludge for Nutrient Recovery Final Report. Available online: https://cordis.europa.eu/project/id/16079 (accessed on 14 November 2020).
88. Venglovsky, J.; Martinez, J.; Placha, I. Hygienic and Ecological Risks Connected with Utilization of Animal Manures and Biosolids in Agriculture. *Livest. Sci.* **2006**, *102*, 197–203. [CrossRef]
89. Collivignarelli, M.; Abbà, A.; Frattarola, A.; Carnevale Miino, M.; Padovani, S.; Katsoyiannis, I.; Torretta, V. Legislation for the Reuse of Biosolids on Agricultural Land in Europe: Overview. *Sustainability* **2019**, *11*, 6015. [CrossRef]
90. Troschinetz, A.M.; Mihelcic, J.R. Sustainable Recycling of Municipal Solid Waste in Developing Countries. *J. Waste Manag.* **2009**, *29*, 915–923. [CrossRef] [PubMed]
91. di Bella, V.; Ali, M.; Vaccari, M. Constraints to Healthcare Waste Treatment in Low-Income Countries–A Case Study from Somaliland. *Waste Manag. Res.* **2012**, *30*, 572–575. [CrossRef] [PubMed]
92. Tai, J.; Zhang, W.; Che, Y.; Feng, D. Municipal Solid Waste Source-Separated Collection in China: A Comparative Analysis. *J. Waste Manag.* **2011**, *31*, 1673–1682. [CrossRef]
93. Gontard, N.; Sonesson, U.; Birkved, M.; Majone, M.; Bolzonella, D.; Celli, A.; Angellier-Coussy, H.; Jang, G.-W.; Verniquet, A.; Broeze, J.; et al. A Research Challenge Vision Regarding Management of Agricultural Waste in a Circular Bio-Based Economy. *Crit. Rev. Environ. Sci. Technol.* **2018**, *48*, 614–654. [CrossRef]
94. UN Environment. *Waste Management Outlook for Latin America and the Caribbean*; Savino, A., Ed.; United Nations Environment Programme: Panama City, Panama, 2018; ISBN 978-92-807-3714-1.
95. Hoomweg, D.; Giannelli, N. *Managing Municipal Solid Waste in Latin America and the Caribbean: Integrating the Private Sector, Harnessing Incentives*; Gridlines; World Bank: Washington, DC, USA, 2007; pp. 1–4.
96. Hettiarachchi, H.; Ryu, S.; Caucci, S.; Silva, R. Municipal Solid Waste Management in Latin America and the Caribbean: Issues and Potential Solutions from the Governance Perspective. *Recycling* **2018**, *3*, 19. [CrossRef]
97. Margallo, M.; Ziegler-Rodriguez, K.; Vázquez-Rowe, I.; Aldaco, R.; Irabien, Á.; Kahhat, R. Enhancing Waste Management Strategies in Latin America under a Holistic Environmental Assessment Perspective: A Review for Policy Support. *Sci. Total Environ.* **2019**, *689*, 1255–1275. [CrossRef]
98. MADS. *Se Reglamenta El Uso de Las Aguas Residuales Y Se Adoptan Otras Disposiciones, Resolución 1256*; Ministerio de Ambiente y Desarrollo Sostenible: Bogotá, Colombia, 2021; pp. 1–6.
99. Bittencourt, S.; Aisse, M.M.; Serrat, B.M. Gestão Do Uso Agrícola Do Lodo de Esgoto: Estudo de Caso Do Estado Do Paraná, Brasil. *Eng. Sanit. E Ambient.* **2017**, *22*, 1129–1139. [CrossRef]
100. Vélez Zuluaga, J.A. Los Biosólidos: ¿Una Solución o Un Problema? *Prod. Más Limpia* **2007**, *2*, 57–71.
101. Spinosa, L. *Status and Perspectives of Sludge Management*; IWA Publishing: Kota Banjarmasin, Indonesia, 2007; pp. 103–108.
102. Salazar Espitia, J. Guía Metodológica Para El Manejo Y Aprovechamiento de Biosólidos en Colombia. Master's Thesis, Universidad Nacional de Colombia, Manizales, Colombia, 2019.
103. Rivera León, F.A. Los Sistemas de Apoyo en La Toma de Decisiones. *Gest. Terc. Milen.* **2014**, *17*, 69–75. [CrossRef]
104. Palma-Heredia, D.; Poch, M.; Cugueró-Escofet, M.À. Implementation of a Decision Support System for Sewage Sludge Management. *Sustainability* **2020**, *12*, 9389. [CrossRef]
105. Horn, A.L.; Düring, R.-A.; Gäth, S. Comparison of Decision Support Systems for an Optimised Application of Compost and Sewage Sludge on Agricultural Land Based on Heavy Metal Accumulation in Soil. *Sci. Total Environ.* **2003**, *311*, 35–48. [CrossRef]
106. van Dam, K.H.; Feng, B.; Wang, X.; Guo, M.; Shah, N.; Passmore, S. Model-Based Decision-Support for Waste-to-Energy Pathways in New South Wales, Australia. *Comput. Aided Chem. Eng.* **2019**, *46*, 1765–1770. [CrossRef]
107. Koukoulakis, P.H.; Kyritsis, S.S.; Kalavrouziotis, I.K. The Contribution of Decision Support System (DSS) to the Approach of the Safe Wastewater and Biosolid Reuse. In *Wastewater and Biosolids Management*; 2020.
108. Castillo, A.; Porro, J.; Garrido-Baserba, M.; Rosso, D.; Renzi, D.; Fatone, F.; Gómez, V.; Comas, J.; Poch, M. Validation of a Decision Support Tool for Wastewater Treatment Selection. *J. Environ. Manag.* **2016**, *184*, 409–418. [CrossRef]
109. Vlachokostas, C.; Achillas, C.; Diamantis, V.; Michailidou, A.V.; Baginetas, K.; Aidonis, D. Supporting Decision Making to Achieve Circularity via A Biodegradable Waste-to-Bioenergy and Compost Facility. *J. Environ. Manag.* **2021**, *285*, 112215. [CrossRef]

Article

Characterization and Planning of Household Waste Management: A Case Study from the MENA Region

Feriel Kheira Kebaili [1], Amel Baziz-Berkani [1], Hani Amir Aouissi [2,*], Florin-Constantin Mihai [3], Moustafa Houda [4], Mostefa Ababsa [2], Marc Azab [4], Alexandru-Ionut Petrisor [5,6,7] and Christine Fürst [8,9]

1. Laboratoire de Recherche et d'Etude en Aménagement et Urbanisme (LREAU), USTHB, Algiers 16000, Algeria; kebailifer.94@gmail.com (F.K.K.); bazizusthb@yahoo.fr (A.B.-B.)
2. Scientific and Technical Research Center on Arid Regions (CRSTRA), Biskra 07000, Algeria; mostefa.ababsa@gmail.com
3. CERNESIM Center, Department of Exact Sciences and Natural Sciences, Institute of Interdisciplinary Research, "Alexandru Ioan Cuza" University of Iasi, 700506 Iasi, Romania; mihai.florinconstantin@gmail.com
4. College of Engineering and Technology, American University of the Middle East, Kuwait; moustafa.houda@aum.edu.kw (M.H.); marc.azab@aum.edu.kw (M.A.)
5. Doctoral School of Urban Planning, Ion Mincu University of Architecture and Urbanism, 010014 Bucharest, Romania; alexandru.petrisor@uauim.ro
6. National Institute for Research and Development in Tourism, 50741 Bucharest, Romania
7. National Institute for Research and Development in Constructions, Urbanism and Sustainable Spatial Development URBAN-INCERC, 021652 Bucharest, Romania
8. Department of Sustainable Landscape Development, Institute for Geosciences and Geography, Martin-Luther-University Halle-Wittenberg, Von-Seckendorff-Platz 4, 06120 Halle, Germany; christine.fuerst@geo.uni-halle.de
9. German Centre for Integrative Biodiversity Research (iDiv) Halle-Jena-Leipzig, Puschstr. 4, 04103 Leipzig, Germany
* Correspondence: aouissi.amir@gmail.com; Tel.: +213-6-6238-7144

Abstract: Solid waste management is one of the most important environmental issues worldwide, particularly in MENA countries. The present study was carried out in the city of Algiers, the capital city of Algeria. This urban area is marked by an increase in waste flow combined with a demographic surge. In order to investigate waste production and its drivers, we used both multiple regression and correlation analyses to test this dependence. Geospatial analysis was performed using principal component analysis integrated with GIS in order to look at the spatial distribution of waste management and potential drivers of waste production. The results indicate that household waste management is influenced by drivers related to the size of the settlement and the characteristics of waste management companies ($p \leq 0.05$). The findings also show that none of the sociodemographic variables were found to significantly influence waste production. However, the spatial distribution is influenced by the geographic and sociodemographic characteristics of Algeria at all territorial levels. Algiers is still a landfill-based city in the MENA region, where mixed waste collection prevails in all districts. This study reinforces the importance of expanding source-separated waste collection schemes in order to increase the household waste diversion from landfills and, more importantly, shows how modern tools such as GIS, principal component analysis, and spatial analysis urban planning are useful for monitoring household waste, in line with circular economy principles.

Keywords: household waste; waste management; spatial analysis; urban planning; Algiers

1. Introduction

The world is experiencing an environmental crisis, marked by excessive production of waste, in particular household waste fed by consumerism and the effects of urbanization [1–3]. The most optimistic prediction says that 70% of humans will live in cities by the year 2050 [4]. However, 0.9 billion people lack access to regular waste collection services in urban areas worldwide, particularly in low- and middle-income countries [5]. The expansion of sanitation and waste management infrastructures in urban areas is crucial to achieving sustainable development goals by 2030. Furthermore, the model of a smart city (SC) goes beyond an urban space where information and communication technologies (ICT) are applied [6]. In fact, the aim of a smart city is to improve the performance and the quality of urban services (energy, transportation, and other infrastructures), intending to reduce costs, resource energy consumption, and wastage. Smart city environments evolve to improve the quality of life of citizens and the operational efficiency of complex urban systems [7].

Solid waste management (SWM) is one of the major challenges faced by smart cities (and cities, in general), especially due to population growth and urbanization [8]. The world's annual waste generation is actually 2.01 billion tons and 0.74 kg/person per day (33% of that is not managed in an environmentally safe manner) and is expected to increase to 3.40 billion tons in 2050 [9]. High-income countries generate about 34% of the world's waste, while the total quantity generated in low-income countries is expected to increase more than three times by 2050. The Middle East and North Africa region produces 6% of the world's waste. In addition, given the fastest growth of cities in the MENA regions, it is expected that by 2050, total waste generation is expected to increase twice (or thrice) [10]. Consequently, there is an urgent need for more efficient solid waste management in cities. SWM is also a major concern for national (and municipal) governments in order to preserve natural resources and the environment and protect human health. The emerging smart technologies already used in waste management, such as robot recyclers, internet of things (IoT), self-driving trucks, and waste level sensors, can further optimize the collection and treatment operations in the largest cities around the globe [11].

Generally, waste characterization and generation are among the most important factors to consider when selecting the most appropriate collection and treatment methods and also the final disposal [12]. This is especially the case in countries in the process of urbanization experiencing increasing population and lifestyle changes under the impact of massive migration from rural to urban zones, which leads to a considerable increase in urban waste generation.

Solid waste and household waste prediction can be conducted at different temporal (e.g., week, month, or year) or geographic (e.g., municipal, regional, national) levels. Country-level studies use previously collected data on the total annual waste amount, waste types, or socioeconomic data, which they often make available to international associations [13]. The applicability of such projections is highly dependent on model assumptions and the quality of data acquired [14]. Solid waste clustering enables discovering differences and similarities among analyzed regions or countries with respect to waste management. Moreover, they also allow for determining the relationships between clusters and demographic, socioeconomic, and waste generation characteristics. However, these inherent structures are difficult to observe in the original datasets because of the multidimensional nature of data [15].

Countries of MENA regions are still reliant on landfills as the main disposal route for household waste, but efforts are made to estimate source-separation collection schemes and recycling practices [16].

Algeria is the largest nation in Africa and the Arab world by area [17]. Its population of 44 million has a heterogeneous density (high population in the north and low in the arid regions), resulting in a high difference in waste generation between urbanized cities and the other regions. The waste management sector in Algeria is insufficiently regulated [18]. Daily and annual waste production can be estimated from the average rate of waste produced per person per day. In fact, an Algerian produces an average of 0.81 kg of household waste per day. This production continues to increase, exceeding 8.5 million tons in 2021 [16]. Algeria is marked by a significant increase in waste flows combined with a demographic surge and a saturation of urban waste management infrastructures [19]. In this context, several uncontrolled dumps have appeared on the Algerian territory, generating direct negative impacts on the environment through the creation of pollution, posing major risks to human health [20]. Hence, the Algerian government is constantly looking to adopt technological solutions in an attempt to treat household waste. These different solutions must protect people and communities, minimize negative impacts on the environment, and allow for efficient and less expensive treatment of waste [21,22].

Studies investigating urbanization, sustainable development, and, more specifically, waste management in North Africa (especially in Algeria, and even more precisely in Algiers) are scarce, incomplete, or even lacking [23,24]. The few existing studies are old and provide very few insights [25,26]. In the current context, studies addressing these themes could be of considerable importance [27].

Although several tools have been implemented for the planning and development of Algiers, the capital city of Algeria, none have been successful. This clearly reflects the failure of urban planning in the capital, which highlights the different management problems encountered by the city [28].

The aim of this study is to investigate and identify the elements and generating factors of the excessive production of household waste at the wilaya of Algiers, through a statistical and geospatial approach, as well as the management practices and their approaches to adopt a better strategy for the optimization of waste management. We hypothesize, according to our knowledge of the study area, that the increase in household waste is linked not only to population growth, but also to other factors such as education, standard of living, urban structure, recycling infrastructure, collection system, and economic and social development. This study will attempt to verify the aforementioned hypothesis, filling in a gap in the existing literature concerning the management of household waste in the study area.

2. Materials and Methods

2.1. Case Study Area

The present study was carried out in the city of Algiers, which is the political, administrative, and economic capital of the country. It is the seat of all central, political, and social institutions, major economic and financial establishments, major decision-making centers, and diplomatic representations.

The "wilaya" (administrative division) of Algiers is located in the north at the following coordinates: 36°46′34″ N, 3°03′36″ E. It occupies a geostrategic position. It spans more than 808.89 km^2 and is limited by the Mediterranean Sea to the north, the wilayas of Blida in the south, Tipaza in the west, and Boumerdès in the east.

According to the 2008 RGPH, the population of Algiers is about 2,987,160 inhabitants. In 2021, the population was estimated at 3.4 million inhabitants [29]. The population density amounts to ~4450 inhabitants/km^2. The wilaya of Algiers is made up of thirteen daïras (administrative districts), each comprising several "communes" (municipalities), for a total of fifty-seven communes.

The study area consists of the 57 communes composing the wilaya of Algier (see Figure 1).

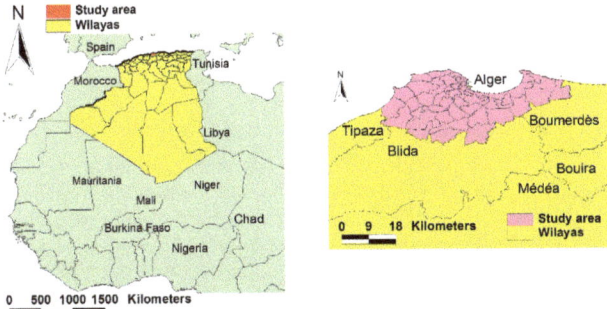

Figure 1. Location of the case study. The images present the location in an international context (**left**) and in detail (**right**).

2.2. Data

The data used in this research come from many sources, including the two companies responsible for waste management in Algiers. Data were collected from the technical department of their general management. The two companies are EPIC EXTRANET, situated in Bab Ezzouar-Algiers, and responsible for 31 communes, and EPIC NETCOM, situated in Mohamed Nail-Algiers, and responsible for 26 communes.

Figure 2 displays the companies responsible for waste management in each commune within the area analyzed in the study.

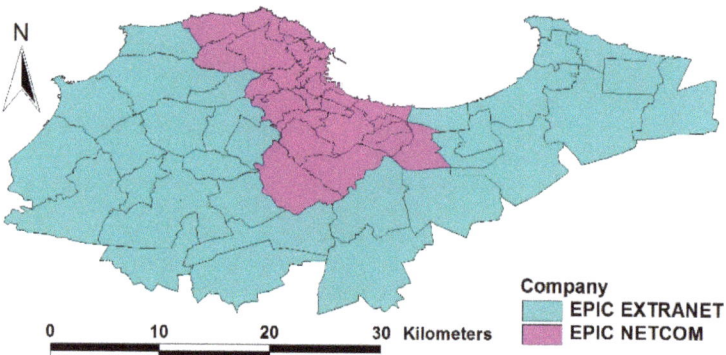

Figure 2. Spatial distribution of companies responsible for waste management across the study area.

Other data were acquired from the National Office of Statistics (ONS) situated in Ruisseau, Algiers, and the National Waste Agency (AND) situated in Hamma, Algiers.

2.3. Variables

Some of the variables used in this work are similar to those used in studies addressing municipal waste management issues in other countries [30–32]. However, some variables, such as the role of the environmental awareness group (EAG), availability of facilities and infrastructure (FI), and budget availability (BA), used in other studies, were excluded either because they were only present in one company, or were obsolete. Consequently, the selected variables were:

- Area (A): total area of the municipality.
- Population (P): total number of inhabitants in the municipality.
- Share of people with at least a college degree (C): number of individuals with medium or higher education.
- Share of active population (AP): part of the population engaged in work or studies.
- Number of garbage bins (G): Total number of garbage bins owned by the concerned company.
- Collection routes (R): predefined waste collection points.
- Number of staff in waste management companies (S): human resources available at the company.
- Company managing waste collection (CO): Extranet or Netcom.

Data provided by waste operators and the aforementioned variables were further processed using spatial analysis and GIS techniques. The scale of analysis, i.e., district level, reveals the local variations in household waste generation within urban administrative borders. The microlevel spatial analysis of such variables has benefits for decision makers, such as adapting urban waste management policies and infrastructure demands. Big data and machine learning could further predict waste generation flow at the microlevel, including building-level data [33]. Reliable historical waste-related data lead to better future predictions of household waste at the district or urban levels, improving the planning of waste management infrastructure by adopting feasible circular economy targets [34,35]. On the other hand, the pro-environmental behavior of residents could be used as an additional variable to examine waste generation patterns at the city level [36].

2.4. Quantitative Analyses

The quantitative analyses were based on statistical approaches used to look at the variables influencing the production of waste, especially the total amount of waste. Geostatistical analyses were also used in order to look at the spatial distribution of the production of waste and its potential drivers. For this purpose, multiple regression analyses were used to look at the possible drivers influencing the total amount of waste, using two models: a full model, looking at the simultaneous influence of all drivers, and a restricted model, obtained from the previous one using backward elimination, containing all drivers with a simultaneous statistically significant influence on the total amount of waste.

The variables included in these models were derived from the data obtained from the statistical office or based on the row data. The new variables were the average production of waste and the number of collection routes, obtained by averaging in each case the values for the eight months included in the row data. Education level was synthesized in only one index that presents the share of those with at least a college degree among all people. Another driver, the employment structure, was presented by the activity rate in the studied area.

The geospatial methods were based on a modified version, "principal component analysis integrated with GIS", a method developed for the identification of "hotspots" that are at the core of intervention policies [37]. This method has been also used in other studies due to its ability to compare the empirical dimension of the factors with the three dimensions [38]. Essentially, the method consists of (1) running a principal component analysis to identify the key variables that can be used to underline the spatial differentiation between different administrative units through differences between their values across the analyzed space, (2) creating an index using the percentage of variation explained by each variable as its weight, and (3) mapping the distribution of the index. The difference from previous applications of the method was that instead of looking at the initial eigenvalues and using a threshold value of 1.00 for the total variance explained, resulting in identifying only two principal components, we considered the extraction sums of squared loadings and used a threshold of 5% for the percentage of variance, identifying four principal components.

3. Results

In order to investigate waste production and its drivers, we used two types of analyses, i.e., multiple regression and correlation analyses to test the dependence, and principal component analysis integrated with GIS to look at the spatial distribution.

3.1. Results of Multiple Regression and Correlation Analyses

Multiple regression analysis used two models, a full model and a prediction model, obtained from the full model through backward elimination. Both models were significant overall ($p < 0.0001$). The overall coefficient of correlation, showing the percentage of variation explained by the model, was 0.892 for the full model and 0.888 for the prediction model. The influence of analyzed drivers is presented in Table 1.

The results of correlation analysis are presented in Table 2. The table displays all possible correlations between variables describing waste management and potential drivers, analyzing those significant ($p \leq 0.05$) and those marginally significant ($p \leq 0.1$). For the latest, it is expected that more data would turn them into significant correlations.

Table 1. Results of multiple regression analysis looking at the dependence of waste production (amount) on socioeconomic and territorial drivers: A—area, P—population, C—share of people with at least a college degree, AP—share of active population, G—number of garbage bins, R—collection routes, S—no. of staff in waste management companies, and CO—company managing waste collection. The table presents two models, a "full model", including all drivers, and a "prediction model", including only those significantly influencing waste production when considered simultaneously. **Bold** values indicate variables significant at $p \leq 0.05$.

Variable	Full Model					Prediction Model				
	DF	Type III SS	Mean Square	F Value	p-Value	DF	Type III SS	Mean Square	F Value	p-Value
A	1	1,176,924.116	1,176,924.116	10.01	0.0027	1	1,488,816.135	1,488,816.135	13.24	0.0006
P	1	1,078,283.887	1,078,283.887	9.17	0.0040	1	1,105,096.310	1,105,096.310	9.82	0.0028
C	1	161,008.924	161,008.924	1.37	0.2477					
AP	1	151,823.646	151,823.646	1.29	0.2615					
G	1	24,613.729	24,613.729	0.21	0.6494					
R	1	1,569,441.511	1,569,441.511	13.35	0.0006	1	2,053,790.064	2,053,790.064	18.26	<0.0001
S	1	895,123.090	895,123.090	7.61	0.0082	1	1,126,868.026	1,126,868.026	10.02	0.0026
CO	1	2075.647	2075.647	0.02	0.8949					

Table 2. Correlation between the variables describing waste management and potential drivers: W—amount of waste, A—area, P—population, C—share of people with at least a college degree, AP—share of active population, G—number of garbage bins, R—collection routes, and S—no. of staff in waste management companies. For each correlation, the table displays the coefficient of determination r and its corresponding p-value below. **Bold** values indicate correlations significant at $p \leq 0.05$, and *italic* values correlations significant at $p \leq 0.1$.

.		W	A	P	C	AP	G	R	S
W	r	1.00000	0.53687	0.83112	−0.00077	−0.36286	0.24162	0.82540	0.81927
	p		<0.0001	<0.0001	0.9955	**0.0055**	*0.0702*	**<0.0001**	**<0.0001**
A	r	0.53687	1.00000	0.20285	−0.44136	−0.53232	0.05287	0.26159	0.58813
	p	**<0.0001**		0.1302	**0.0006**	**<0.0001**	0.6961	**0.0493**	**<0.0001**
P	r	0.83112	0.20285	1.00000	0.15169	−0.17607	0.35841	0.82465	0.69092
	p	**<0.0001**	0.1302		0.2600	0.1902	**0.0062**	**<0.0001**	**<0.0001**
C	r	−0.00077	−0.44136	0.15169	1.00000	0.63193	0.13437	0.11128	−0.04608
	p	0.9955	**0.0006**	0.2600		**<0.0001**	0.3190	0.4099	0.7336
AP	r	−0.36286	−0.53232	−0.17607	0.63193	1.00000	−0.14446	−0.24338	−0.30994
	p	**0.0055**	**<0.0001**	0.1902	**<0.0001**		0.2837	*0.0681*	**0.0190**
G	r	0.24162	0.05287	0.35841	0.13437	−0.14446	1.00000	0.24407	0.14963
	p	*0.0702*	0.6961	**0.0062**	0.3190	0.2837		*0.0673*	0.2666
R	r	0.82540	0.26159	0.82465	0.11128	−0.24338	0.24407	1.00000	0.58743
	p	**<0.0001**	**0.0493**	**<0.0001**	0.4099	*0.0681*	*0.0673*		**<0.0001**
S	r	0.81927	0.58813	0.69092	−0.04608	−0.30994	0.14963	0.58743	1.00000
	p	**<0.0001**	**<0.0001**	**<0.0001**	0.7336	**0.0190**	0.2666	**<0.0001**	

3.2. Results of Geospatial Analyses

The results of geospatial analysis are based on a principal component analysis, used to look at the territorial variables describing waste management and potential drivers of waste production. The results of the analysis are displayed in Tables 3 and 4. Table 3 shows the principal components and the percentage of variation explained by each; out of these, we considered the extraction sums of squared loadings and used a threshold of 5% for the percentage of variance to identify the principal components with a significant influence on the territorial distribution. Table 4 serves for identifying them with the variable most correlated to each one, as indicated by the highest value of the coefficient of determination.

The results presented in Table 3 indicate that the first four components, accounting altogether for over 90% of the total variation (in terms of the extraction sums of squared loadings), and identified in Table 4 with the amount of waste, share of people with at least a college degree, number of garbage bins, and area of the commune, can be used to analyze the territorial differences across the study area. This was achieved by mapping the spatial distribution of each variable (see Figures 3–6) and building an aggregated index weighting each variable based on the percentage of total variance explained (column "Extraction Sums of Squared Loadings" in Table 3) and mapping its spatial distribution (see Figure 7).

Table 3. Results of the principal component analysis showing the principal components explaining the territorial differences of communes with respect to waste management and potential drivers.

Component	Initial Eigenvalues			Extraction Sums of Squared Loadings		
	Total	% of Variance	Cumulative %	Total	% of Variance	Cumulative %
1	3.823	47.788	47.788	3.823	47.788	47.788
2	1.912	23.902	71.690	1.912	23.902	71.690
3	0.936	11.705	83.394	0.936	11.705	83.394
4	0.589	7.357	90.752	0.589	7.357	90.752
5	0.291	3.640	94.391	0.291	3.640	94.391
6	0.261	3.260	97.651	0.261	3.260	97.651
7	0.109	1.365	99.016	0.109	1.365	99.016
8	0.079	0.984	100.000	0.079	0.984	100.000

Table 4. Results of the principal component analysis identifying the four principal components explaining over 5% of the variance (column "Extraction Sums of Squared Loadings" in Table 3) with the corresponding variables. The correspondence is indicated by the highest value of the coefficient of determination in each column (1–4), underlined using a **Bold** font.

Variable	Component			
	1	2	3	4
Area	0.620	−0.559	0.124	**0.461**
Population	0.846	0.396	−0.015	−0.196
Share of people with at least a college degree	−0.130	**0.889**	0.088	0.227
Share of active population	−0.505	0.688	0.273	0.251
Garbage bins	0.338	0.263	−**0.872**	0.220
Amount of waste	**0.953**	0.121	0.128	−0.003
Collection routes	0.830	0.311	0.063	−0.321
No. of staff in waste management companies	0.861	0.014	0.239	0.268

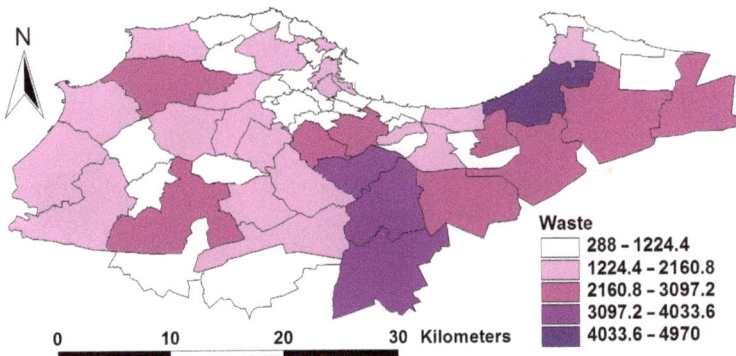

Figure 3. Spatial distribution of the production of waste across the study area. The distribution is based on creating five equal intervals based on the amount of waste generated in each unit (in metric tons).

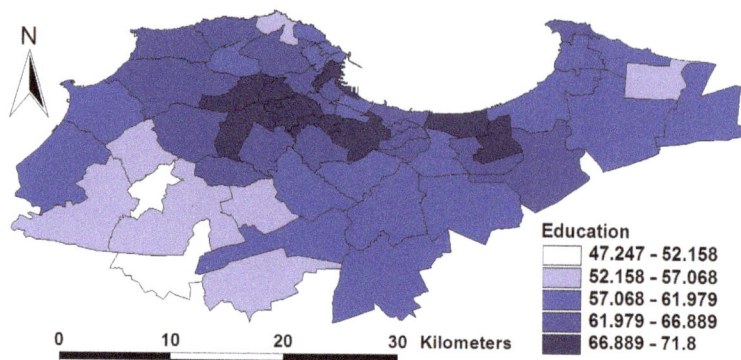

Figure 4. Spatial distribution of the share of people with at least a college degree across the study area. The distribution is based on creating five equal intervals based on the share of people with at least a college degree from the total population.

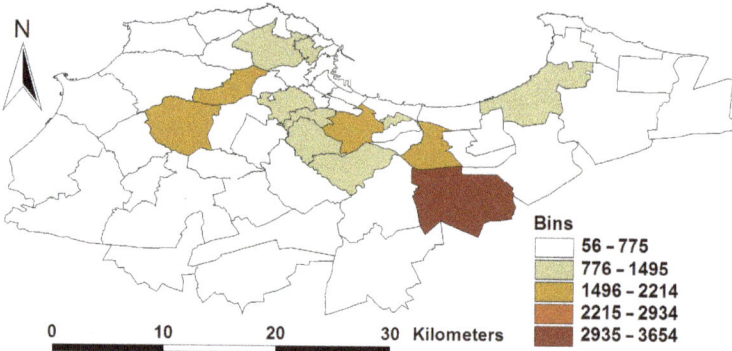

Figure 5. Spatial distribution of the number of garbage bins across the study area. The distribution is based on creating five equal intervals.

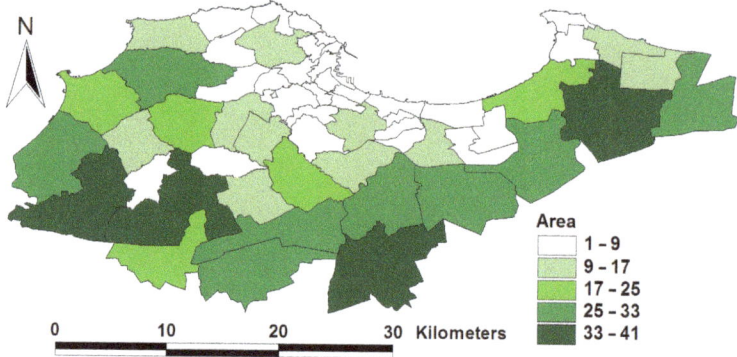

Figure 6. Spatial distribution of the total commune area across the study area. The distribution is based on creating five equal intervals for the total area of communes, in km^2.

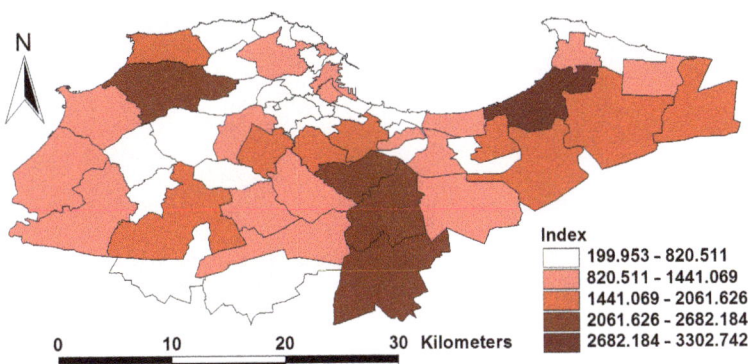

Figure 7. Spatial distribution of the aggregated index across the study area. The distribution is based on creating five equal intervals for aggregated index, built using the four principal components explaining over 90% of the total variation (in terms of the extraction sums of squared loadings—Table 3) weighted by the percentage of variation explained (Table 3).

4. Discussion

The results of multiple regression analysis indicate that waste management is influenced by drivers related to the size of the settlement, specifically area ($p = 0.0006$) and population ($p = 0.0028$). Waste management is also influenced by the characteristics of WM companies, represented by the number of collection routes ($p < 0.0001$) and number of employees ($p = 0.0026$). None of the sociodemographic variables such as the level of education (people with at least college level, active population) or employment rate were found to significantly influence the production of waste. Similar to other studies, population and size of urban area are found to be key drivers in the waste generation rates [30,34].

The spatial distributions are influenced by the geographic and sociodemographic characteristics of Algeria, and the influence is seen at all territorial levels: wilayas within the country, communes within the wilayas. The size of communes increases from north to south (see Figure 5), because in Algeria, the north corresponds to the Mediterranean shore, which offers, due to its climate, better living conditions and is more populated. The south corresponds to the desert areas and is less populated. As a result, the size balances the communes in terms of population (see Figure 3), i.e., communes are larger in less populous areas. The same geographical settings affect the spatial distribution of education; the north concentrates large cities with strong universities, while the southern areas offer fewer educational opportunities.

This paper reveals the importance of urban district-level variation in terms of waste production, waste bin coverage, and sociodemographic features, as in other studies [31]. At the microlevel, house size, districts, and employment category were found key drivers in waste generation for the Accra region of Ghana [32] and family size in Robe town, Ethiopia [39].

Therefore, other factors could be further examined in future research such as household/family size. The employment category seems not to play a significant role in explaining waste production in Algiers city. Furthermore, spatial analysis using GIS tools provides better monitoring alternatives of waste indicators to improve the collection efficiency. This is important to prevent illegal disposal activities in emerging cities [40].

Compared to other African countries, Algeria is better covered by basic public utilities in both urban and rural regions [5,41]. The population of Algiers city and wilaya is connected to waste collection services, but a traditional waste management system based on mixed waste collection and landfill of waste prevails in the capital city. The household waste collected by two main waste operators (NETCOM and EXTRANET) is disposed of in two different conventional landfill sites (CET HAMICI and CET of Corso). However, the issue of uncontrolled household waste disposal practices is not solved, contributing to environmental pollution [42]. The illegal dumping of household waste remains a significant environmental threat in African cities [39].

The urban waste generation rate is estimated to be at 0.8 kg.inhab.day-1 compared to 0.6 kg.inhab.day-1 in rural areas at the country level, while the recycling rate is around 7%, composting 1%, and the rest of municipal waste flow is disposed of in sanitary landfills or dumpsites [43]. The waste generation rate varies among regions of Algeria with significant disparities between north and south regions, but the highest waste generation rate is around 0.95 kg.inhab.day-1 in Algiers [20]. Across estimation of household waste generated by Algiers city is around 500,000 tons per semester, approximately 1 million tons per year [44]. Therefore, this study provides a better picture of household waste flows using spatial analysis at the district level. Such spatial analyses at this scale are needed by decision makers aiming to improve the current waste management practices. However, experimental analyses based on field data regarding urban waste generation rates on a per capita basis and municipal waste composition in different districts of Algiers city and wilaya are required to provide a comprehensive baseline of household waste characterization for the study area and model future predictions of waste flows. The frequency of waste collection schemes is a daily regime and in some periods of the year the regime is twice a day (e.g., summer season for coastal communes and the two days of the "Aid el Adha" celebration). During the month of "Ramadan", the waste collection regime becomes nightly. These religious events could significantly increase the urban waste generation rates, putting additional pressure on waste collection schemes [45]. This issue could also be investigated in Algiers city, in comparison with other urban centers of the country.

There are no source-separation collection schemes, but some recyclable waste is recovered from residual waste at the sorting station located near the CET landfill site. Additionally, the informal sector collects some dry recyclable waste (plastics and metals) for recycling purposes. The informal sector plays a critical role in African cities in recovering recyclable waste from dumpsites associated with the underdevelopment of municipal waste management infrastructure [46]. Most private economic agents involved in waste recovery and recycling operations are found in Algiers and Boumerdes according to the National Waste Agency [47]. Metals and scraps of iron are often recovered at the source, and these fractions are less prone to landfilling compared to the organic waste fraction [21].

NETCOM introduced seven separate collection sites for paper/cardboard, glass, organic waste, and bread [48]. Besides waste collection operations, the transportation of household waste seems to have a high ecological footprint, and landfills are not equipped with biogas installations in Algiers [42]. The rate of recycling household waste is unknown, and computation of waste statistics at the city level should be compulsory. At the country level, the latest estimation of the recycling rate was around 9.83% in 2020 [47]. There are no composting facilities in Algiers despite the fact that most residual bins contain organic waste (54.4%), followed by plastics (16.5%) and paper/cardboard (13.4%) [48]. This is in line with the last household waste characterization from April 2018 to March 2019, where the organic fraction represented 63%, followed by plastics (15.2%) [49]. However, this characterization did not include Algiers. Therefore, such studies need to be further developed in the capital city and related wilayas. Preliminary findings suggest that composting facilities fed by source separation of organic waste should be a future investment priority besides material recycling facilities combined with separate collection schemes extended to all districts in the study area. Composting practices are limited across MENA countries despite the high share of organic waste in household waste flows [15].

The recovery of recyclable waste is very low because sorting stations receive mixed household waste instead of clean source-separated recyclable wastes (paper/cardboard, glass, plastic, and metals/aluminum cans). Without proper waste management infrastructure separating clean organic wastes from dry recyclable waste, the target of recycling and composting more than 50% of household and similar waste by 2035, set by the Waste Management Strategy in Algeria [50], will be an impossible target to fulfill based on the current situation in Algeria. In fact, the African Circular Economy Alliance argues the key role of organic waste and plastic packaging diversion from landfills having as members Cote d'Ivoire, Ghana, Nigeria, Rwanda, and South Africa [51]. A similar alliance could be initiated in the MENA region, where Algeria could be part of it or adhere to the existing ACEA to stimulate transition towards a sustainable waste management system by 2035 through international cooperation, exchanging of know-how, and adopting best practices in the region.

5. Conclusions

This study reveals that GIS and spatial statistics at the district level are useful tools for the assessment and monitoring of household waste flow in large urban areas such as North African capital cities. However, this approach depends on the availability and quality of waste statistics provided by waste operators. The results of multiple regression analysis indicate that waste management is influenced by drivers related to the size of the settlement, i.e., area ($p = 0.0006$) and population ($p = 0.0028$), and characteristics of the waste management companies, i.e., number of collection routes ($p < 0.0001$), and number of employees ($p = 0.0026$). Sociodemographic variables such as the level of education and the employment rate were found to have no significant influence on the production of waste. Algiers is still a landfill-based city in the MENA region, where mixed waste collection prevails in all districts. The expansion of source-separated waste collection schemes is compulsory at least for organic waste and dry recyclables (plastic, metal, paper cardboard, and glass) that must be further treated in composting and sorting facilities to increase the household waste diversion from landfills in line with circular economy principles.

From a methodological perspective, our study proved that a combination of multiple regression analysis and principal component analysis is efficient to describe and understand waste production; the first method is useful in detecting the relevant drivers of waste production and the second in contextualizing them spatially. However, the reliability of our methodology depends at large on the availability of waste statistics provided by waste operators, which can constitute an important challenge in North Africa. Future studies can use more relevant variables to start with, if data are available. In this regard, data on more socioeconomic variables could enhance the results of future studies.

Author Contributions: Conceptualization, A.-I.P., H.A.A. and C.F.; methodology, A.-I.P.; software, M.A. (Mostefa Ababsa); validation, A.-I.P., A.B.-B. and M.H.; formal analysis, A.-I.P.; investigation, F.K.K.; resources, A.B.-B. and H.A.A.; data curation, M.A. (Mostefa Ababsa), F.-C.M. and C.F.; writing—original draft preparation, A.-I.P., F.-C.M. and H.A.A.; writing—review and editing, F.K.K., A.-I.P., F.-C.M., M.H., M.A. (Marc Azab) and H.A.A.; visualization, A.-I.P. supervision, A.B.-B., A.-I.P., F.-C.M. and C.F.; project administration, A.B.-B.; funding acquisition, M.H., M.A. (Marc Azab) and H.A.A. All authors have read and agreed to the published version of the manuscript.

Funding: This study was supported by the DGRSDT and the MESRS.

Institutional Review Board Statement: Not applicable.

Informed Consent Statement: Not applicable.

Data Availability Statement: The data presented in this study are collected within a research project and can be made available on request from the corresponding author.

Acknowledgments: Many thanks are addressed to the technical services of both EXTRANET and NETCOM societies for sharing their database. Many thanks are also addressed to the general directions of the Algerian National Office of Statistics (ONS) and the Algerian National Waste Agency (AND) for providing the data used in this study. Considerable appreciation is addressed to the DGRSDT (Directorate-General of Scientific Research and Technological Development of Algeria) and the MESRS (Ministry of Higher Education and Scientific Research of Algeria) for their valuable support.

Conflicts of Interest: The authors declare no conflict of interest.

References

1. Yoada, R.M.; Chirawurah, D.; Adongo, P.B. Domestic Waste Disposal Practice and Perceptions of Private Sector Waste Management in Urban Accra. *BMC Public Health* **2014**, *14*, 697. [CrossRef] [PubMed]
2. Ferronato, N.; Torretta, V. Waste Mismanagement in Developing Countries: A Review of Global Issues. *Int. J. Environ. Res. Public Health* **2019**, *16*, 1060. [CrossRef] [PubMed]
3. Valceanu, D.-G.; Suditu, B.; Petrisor, A.-I. Romanian Technological Risk Objectives (SEVESO). Effects on Land Use and Territorial Planning. *Carpathian J. Earth Environ. Sci.* **2015**, *10*, 203–209.
4. Ritchie, H.; Roser, M. Urbanization. *Our World in Data* **2018**. Available online: https://ourworldindata.org/urbanization (accessed on 4 February 2022).
5. Mihai, F.-C. One Global Map but Different Worlds: Worldwide Survey of Human Access to Basic Utilities. *Hum. Ecol.* **2017**, *45*, 425–429. [CrossRef]
6. Yeh, H. The Effects of Successful ICT-Based Smart City Services: From Citizens' Perspectives. *Gov. Inf. Q.* **2017**, *34*, 556–565. [CrossRef]
7. Voordijk, H.; Dorrestijn, S. Smart City Technologies and Figures of Technical Mediation. *Urban Res. Pract.* **2021**, *14*, 1–26. [CrossRef]
8. Gutberlet, J. Waste in the City: Challenges and Opportunities for Urban Agglomerations. In *Urban Agglomeration*; InTech: London, UK, 2018.
9. Wowrzeczka, B. City of Waste—Importance of Scale. *Sustainability* **2021**, *13*, 3909. [CrossRef]
10. Kaza, S.; Lisa, Y.; Bhada-Tata, P.; Van Woerden, F. *What a Waste 2.0: A Global Snapshot of Solid Waste Management to 2050*; World Bank: Washington, DC, USA, 2018; Available online: https://openknowledge.worldbank.org/handle/10986/30317 (accessed on 5 February 2022).
11. Kumar, S.; Smith, S.R.; Fowler, G.; Velis, C.; Kumar, S.J.; Arya, S.; Rena; Kumar, R.; Cheeseman, C. Challenges and Opportunities Associated with Waste Management in India. *R. Soc. Open Sci.* **2017**, *4*, 160764. [CrossRef]
12. Phuong, N.; Yabar, H.; Mizunoya, T. Characterization and Analysis of Household Solid Waste Composition to Identify the Optimal Waste Management Method: A Case Study in Hanoi City, Vietnam. *Earth* **2021**, *2*, 1046–1058. [CrossRef]
13. Dunkel, J.; Dominguez, D.; Borzdynski, Ó.G.; Sánchez, Á. Solid Waste Analysis Using Open-Access Socio-Economic Data. *Sustainability* **2022**, *14*, 1233. [CrossRef]
14. Cubillos, M.; Wulff, J.N.; Wøhlk, S. A Multilevel Bayesian Framework for Predicting Municipal Waste Generation Rates. *Waste Manag.* **2021**, *127*, 90–100. [CrossRef] [PubMed]
15. National Research Council. *Prudent Practices in the Laboratory: Handling and Management of Chemical Hazards, Updated Version*; National Academies Press: Washington, DC, USA, 2011; p. 360.
16. Hemidat, S.; Achouri, O.; Fels, L.E.; Elagroudy, S.; Hafidi, M.; Chaouki, B.; Ahmed, M.; Hodgkinson, I.; Guo, J. Solid Waste Management in the Context of a Circular Economy in the MENA Region. *Sustainability* **2022**, *14*, 480. [CrossRef]
17. Worldmeter 2015. Available online: https://www.worldometers.info/population/largest-cities-in-the-world/ (accessed on 2 February 2022).
18. Stambouli, A.B. Algerian Renewable Energy Assessment: The Challenge of Sustainability. *Energy Policy* **2011**, *39*, 4507–4519. [CrossRef]
19. Mohamed, K.; Amina, M.-S.; Mouaz, M.B.E.; Zihad, B.; Wafa, R. The Impact of the Coronavirus Pandemic on the Household Waste Flow during the Containment Period. *Environ. Anal. Health Toxicol.* **2021**, *36*, e2021011. [CrossRef]
20. Abdelkader, O.; Ahmadouche, B. The Problem of Municipal Solid Waste Management in Algeria. *J. New Econ.* **2021**, *12*, 118–131.
21. Okkacha, Y.; Abderrahmane, Y.; Hassiba, B. Municipal Waste Management in the Algerian High Plateaus. *Energy Procedia* **2014**, *50*, 662–669. [CrossRef]
22. Kouloughli, S.; Kanfoud, S. Municipal Solid Waste Management in Constantine, Algeria. *J. Geosci. Environ. Prot.* **2017**, *05*, 85–93. [CrossRef]
23. Guerzou, M.; Aouissi, H.A.; Guerzou, A.; Burlakovs, J.; Doumandji, S.; Krauklis, A.E. From the Beehives: Identification and Comparison of Physicochemical Properties of Algerian Honey. *Resources* **2021**, *10*, 94. [CrossRef]
24. Aouissi, H.A.; Petrişor, A.-I.; Ababsa, M.; Boştenaru-Dan, M.; Tourki, M.; Bouslama, Z. Influence of Land Use on Avian Diversity in North African Urban Environments. *Land* **2021**, *10*, 434. [CrossRef]
25. Bouanini, S. *Assessing the Management of Municipal Solid Waste for Well-Being Fulfillment in Algeria*; University of Eloued: El Oued, Algeria, 2012.
26. Youcef, K.; SWEEP-Net; Lazhari, G. Country Report on the Solid Waste Management in Algeria, The Regional Solid Waste Exchange of Information and Expertise Network in Mashreq and Maghreb Countries SWEEP-Net. 2010. Available online: https://www.scirp.org/(S(351jmbntvnsjt1aadkposzje))/reference/ReferencesPapers.aspx?ReferenceID=1642804 (accessed on 7 January 2022).
27. Louafi, O. The Phenomenon of Mobility, a Development Challenge for the City of Algiers. *J. Contemp. Urban Aff.* **2019**, *3*, 144–155. [CrossRef]
28. Baouni, T. Le Transport Dans Les Stratégies de La Planification Urbaine de l'agglomération d'Alger. *Insaniyat Rev. Algérienne D'anthropologie et de Sci. Soc.* **2009**, *44–45*, 75–95. [CrossRef]

29. World Population Review. Available online: https://worldpopulationreview.com/countries/algeria-population (accessed on 15 February 2022).
30. Pathak, D.R.; Mainali, B.; Abuel-Naga, H.; Angove, M.; Kong, I. Quantification and characterization of the municipal solid waste for sustainable waste management in newly formed municipalities of Nepal. *Waste Manag. Res.* **2020**, *38*, 1007–1018. [CrossRef] [PubMed]
31. Aryampa, S.; Maheshwari, B.; Sabiiti, E.; Bateganya, N.L.; Bukenya, B. Status of Waste Management in the East Afri-can Cities: Understanding the Drivers of Waste Generation, Collection and Disposal and Their Impacts on Kampala City's Sustainability. *Sustainability* **2019**, *11*, 5523. [CrossRef]
32. Chapman-Wardy, C.; Asiedu, L.; Doku-Amponsah, K.; Mettle, F.O. Modeling the Amount of Waste Generated by Households in the Greater Accra Region Using Artificial Neural Networks. *J. Environ. Public Health* **2021**, *2021*, 8622105. [CrossRef]
33. Kontokosta, C.E.; Hong, B.; Johnson, N.E.; Starobin, D. Using machine learning and small area estimation to predict building-level municipal solid waste generation in cities. *Comput. Environ. Urban Syst.* **2018**, *70*, 151–162. Available online: https://www.sciencedirect.com/science/article/pii/S0198971517305859 (accessed on 2 February 2022). [CrossRef]
34. Smejkalová, V.; Šomplák, R.; Nevrlý, V.; Burcin, B.; Kučera, T. Trend forecasting for waste generation with structural break. *J. Clean. Prod.* **2020**, *266*, 121814. [CrossRef]
35. Meza, J.K.S.; Yepes, D.O.; Rodrigo-Ilarri, J.; Cassiraga, E. Predictive analysis of urban waste generation for the city of Bogotá, Colombia, through the implementation of decision trees-based machine learning, support vector machines and artificial neural networks. *Heliyon* **2019**, *5*, e02810. [CrossRef]
36. Sukholthaman, P.; Chanvarasuth, P.; Sharp, A. Analysis of waste generation variables and people's attitudes towards waste management system: A case of Bangkok, Thailand. *J. Mater. Cycles Waste Manag.* **2017**, *19*, 645–656. [CrossRef]
37. Petrișor, A.-I.; Ianoș, I.; Iurea, D.; Văidianu, M.-N. Applications of Principal Component Analysis Integrated with GIS. *Procedia Environ. Sci.* **2012**, *14*, 247–256. [CrossRef]
38. Stoica, I.-V.; Tulla, A.F.; Zamfir, D.; Petrișor, A.-I. Exploring the Urban Strength of Small Towns in Romania. *Soc. Indic. Res.* **2020**, *152*, 843–875. [CrossRef]
39. Erasu, D.; Feye, T.; Kiros, A.; Balew, A. Municipal solid waste generation and disposal in Robe town, Ethio-pia. *J. Air Waste Manag. Assoc.* **2018**, *68*, 1391–1397. Available online: https://www.tandfonline.com/doi/full/10.1080/10962247.2018.1467351 (accessed on 15 February 2022). [CrossRef] [PubMed]
40. Nadeem, K.; Shahzad, S.; Hassan, A.; Usman Younus, M.; Asad Ali Gillani, S.; Farhan, K. Municipal solid waste generation and its compositional assessment for efficient and sustainable infrastructure planning in an inter-mediate city of Pakistan. *Environ. Technol.* **2022**. *(just-accepted)*. [CrossRef] [PubMed]
41. Leveau, C.M.; Aouissi, H.A.; Kebaili, F.K. Spatial diffusion of COVID-19 in Algeria during the third wave. *GeoJournal* **2022**, 1–6. [CrossRef] [PubMed]
42. Akrour, S.; Moore, J.; Grimes, S. Assessment of the Ecological Footprint Associated with Consumer Goods and Waste Management Activities of South Mediterranean Cities: Case of Algiers and Tipaza. *Environ. Sustain. Indic.* **2021**, *12*, 100154. [CrossRef]
43. SweepNet. Report on the Solid Waste Management in ALGERIA. 2014. Available online: https://www.retech-germany.net/fileadmin/retech/05_mediathek/laenderinformationen/Algerien_RA_ANG_WEB_0_Laenderprofile_sweep_net.pdf (accessed on 9 February 2022).
44. Algerian Press Service 2019. Alger: Plus de 500,000 t de Déchets Ménagers Collectés Durant Le 1er Semestre 2019. Available online: https://www.aps.dz/regions/91733-alger-plus-de-500-000-t-de-dechets-menagers-collectes-durant-le-1er-semestre-2019 (accessed on 25 February 2022).
45. Abdulredha, M.; Abdulridha, A.; Shubbar, A.A.; Alkhaddar, R.; Kot, P.; Jordan, D. Estimating municipal solid waste generation from service processions during the Ashura religious event. In *IOP Conference Series: Materials Science and Engineering*; IOP Publishing: Bristol, UK, 2020; Volume 671, p. 012075. Available online: https://iopscience.iop.org/article/10.1088/1757-899X/671/1/012075/ (accessed on 9 February 2022).
46. Ernstson, H.; Lawhon, M.; Makina, A.; Millington, N.; Stokes, K.; Swyngedouw, E. Turning Livelihood to Rubbish? The Politics of Value and Valuation in South Africa's Urban Waste Sector. In *African Cities and Collaborative Futures*; Manchester University Press: Manchester, UK, 2021; p. 96.
47. National Waste Agency. *Algeria Waste Management Status in Algeria*; Deutsche Gesellschaft für Internationale Zusammenarbeit (GIZ) GmbH: Frankfurt, Germany, 2020.
48. Le Tri Sélectif. Available online: https://netcom.dz/tri.html (accessed on 9 February 2022).
49. National Waste Agency. Algeria Caractérisation Des Déchets Ménagers et Assimilés. 2018. Available online: https://and.dz/caracterisation-dechets-menagers-assimiles-algerie-printemps-2018/ (accessed on 9 February 2022).
50. Ghennam, N. Waste Recycling Business in Algeria—Opportunities and Challenges for SME. *Al-Riyada Bus. Econ. J.* **2020**, *6*, 10–22.
51. ACEA. Five Big Bets for the Circular Economy in Africa. 2021. Available online: https://www.weforum.org/reports/five-big-bets-for-the-circular-economy-in-africa-african-circular-economy-alliance (accessed on 9 February 2022).

MDPI
St. Alban-Anlage 66
4052 Basel
Switzerland
Tel. +41 61 683 77 34
Fax +41 61 302 89 18
www.mdpi.com

Sustainability Editorial Office
E-mail: sustainability@mdpi.com
www.mdpi.com/journal/sustainability

www.ingramcontent.com/pod-product-compliance
Lightning Source LLC
LaVergne TN
LVHW070653100526
838202LV00013B/953